THE WILEY BICENTENNIAL—KNOWLEDGE FOR GENERATIONS

*E*ach generation has its unique needs and aspirations. When Charles Wiley first opened his small printing shop in lower Manhattan in 1807, it was a generation of boundless potential searching for an identity. And we were there, helping to define a new American literary tradition. Over half a century later, in the midst of the Second Industrial Revolution, it was a generation focused on building the future. Once again, we were there, supplying the critical scientific, technical, and engineering knowledge that helped frame the world. Throughout the 20th Century, and into the new millennium, nations began to reach out beyond their own borders and a new international community was born. Wiley was there, expanding its operations around the world to enable a global exchange of ideas, opinions, and know-how.

For 200 years, Wiley has been an integral part of each generation's journey, enabling the flow of information and understanding necessary to meet their needs and fulfill their aspirations. Today, bold new technologies are changing the way we live and learn. Wiley will be there, providing you the must-have knowledge you need to imagine new worlds, new possibilities, and new opportunities.

Generations come and go, but you can always count on Wiley to provide you the knowledge you need, when and where you need it!

WILLIAM J. PESCE
PRESIDENT AND CHIEF EXECUTIVE OFFICER

PETER BOOTH WILEY
CHAIRMAN OF THE BOARD

ConcepTests

to accompany

Functions
Modeling
Change
A Preparation for Calculus
Third Edition

by
Eric Connally
Harvard University Extension

Deborah Hughes-Hallett
University of Arizona

Andrew M. Gleason
Harvard University

et al.

John Wiley & Sons, Inc.

This material is based upon work supported by the National Science Foundation under Grant No. DUE-9352905. Opinions expressed are those of the authors and not necessarily those of the Foundation.

ISBN-13 978-0-470-10818-5

Printed in the United States of America

10 9 8 7 6 5 4 3 2 1

Printed and bound by Bind-Rite Graphics / Robbinsville.

CONTENTS

Chapter One

ConcepTests and Answers and Comments for Section 1.1

1. Which one point can be removed from Figure 1.1 to make it the graph of a function?
   (a) A   (b) B   (c) C   (d) D   (e) Any one of the previous

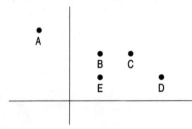

Figure 1.1

ANSWER:
(b)
COMMENT:
Ask about point E.

2. A function can be represented by a
   (a) Verbal description
   (b) Table
   (c) Graph
   (d) Formula
   (e) All of the above

   ANSWER:
   (e)
   COMMENT:
   Ask students to give examples of each.

3. The graph in Figure 1.2 remains the graph of a function when which of the following points are added?
   (a) $(0, 0)$
   (b) $(-100, 1)$
   (c) $(1, -100)$
   (d) All of the above
   (e) $(0, 0)$ and $(-100, 1)$

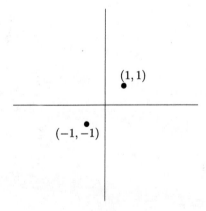

Figure 1.2

ANSWER:
(e)
COMMENT:
Some students automatically connect the points and think that this is the function we are asking about

4. A graph with only one point is always the graph of a function.

   (a) True
   (b) False

   > ANSWER:
   > (a) True

5. A table with only one entry is always the table of values of a function.

   (a) True
   (b) False

   > ANSWER:
   > (a) True

6. A graph with only two points is always the graph of a function.

   (a) True
   (b) False

   > ANSWER:
   > (b) False

7. If $f(x) = 5$ then $f(2) = 10$

   (a) True
   (b) False

   > ANSWER:
   > (b) False
   > COMMENT:
   > You might also ask students if $f(x) \times 2 = 10$.

8. How many letters of the alphabet could be the graph of a function?

   (a) None
   (b) One
   (c) Two
   (d) Most of them
   (e) All of them

   > ANSWER:
   > (c) Only v and w, since the other letters contain two or more points on a vertical line.
   > COMMENT:
   > You may want to give an example to get students started.

ConcepTests and Answers and Comments for Section 1.2

1. Which of the functions shown in Figure 1.3 are increasing?

(a)

(b)

(c)

(d)

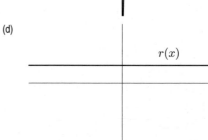

Figure 1.3

ANSWER:

None of them.

COMMENT:

Without a stated interval the function is assumed to be globally increasing. The function $g(x)$ is non-increasing for intervals across the origin.

2. In Figure 1.4, if $h(x) = f(x) - g(x)$ then the rate of change of h is

(a) Not a constant
(b) Positive
(c) Negative
(d) Zero
(e) Undefined

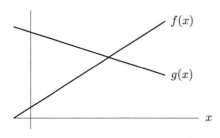

Figure 1.4

ANSWER:

(b)

COMMENT:

Students may need a hint to get them started. Prompt them to think about the rate of change of $f(x)$ and $g(x)$ separately first and then combine them.

3. In Table 1.1 the average rate of change of h on every interval is

    (a) No rate of change
    (b) Positive
    (c) Negative
    (d) Zero

Table 1.1

t	0	24	48	72
$h(t)$	75	75	75	75

ANSWER:

(d)

COMMENT:

Note how (a) and (d) are different.

4. Which table shows an increasing function?

    (a)

t	0	24	48	72
$h(t)$	75	75	75	75

    (b)

t	4	3	2	1
$r(t)$	5	10	15	20

    (c)

t	0	2	4	8
$s(t)$	-25	-20	-15	-10

    (d)

t	0	-24	-48	-72
$m(t)$	5	10	15	20

ANSWER:

(c)

COMMENT:

Ask students why (d) is not a correct answer. Note that the input values are listed in decreasing order.

5. If $T = 0.25n + 100$ then the average rate of change of T with respect to n is

    (a) $\dfrac{\Delta n}{\Delta T}$ (b) $\dfrac{\Delta T}{n}$ (c) $\dfrac{\Delta T}{\Delta n}$ (d) $\dfrac{T}{n}$

ANSWER:

(c)

COMMENT:

New notation is often harder for students than we think. It is worth spending some time to see if they can use it correctly.

6. The average rate of change of a function $T = f(x)$ on an interval is

    (a) $\dfrac{\Delta T}{\Delta x}$ (b) $\dfrac{\text{Rise}}{\text{Run}}$ (c) Slope, $\dfrac{f(b) - f(a)}{b - a}$ (d) All of the above

ANSWER:

(d)

COMMENT:

New notation is often harder for students than we think. It is worth spending some time to see if they can use it correctly.

ConcepTests and Answers and Comments for Section 1.3 ────────────

1. Which table shows data from a linear function?

(a)

t	0	1	2	3
$f(t)$	0	1	4	8

(c)

t	1	0	2	0.1
$h(t)$	3.3	1.2	5.4	1.41

(b)

t	0	1	4	8
$g(t)$	5	10	15	20

(d)

t	−3	0	2	4
$s(t)$	−6	0	6	12

ANSWER:

(c)

COMMENT:

Note that the input values are unevenly spaced and out of order in (c).

2. If points $(1, 3)$, $(3, a)$, and $(b, 11)$ all lie on the same line, then the number of possible values for a and b is

(a) None
(b) One
(c) Two
(d) Infinite

ANSWER:

(d)

COMMENT:

Encourage students to draw a graph.

3. Which of the following points are definitely on the x-axis?

(a) $(0, 1)$
(b) $(b, 0)$
(c) $(0, c)$
(d) $(0, 0)$

ANSWER:

(b) and (d)

COMMENT:

Encourage students to draw a graph.

Use Figure 1.5 to answer Problems 4–7. Use answers:

(a) A
(b) B
(c) C
(d) D
(e) None

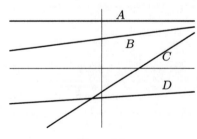

Figure 1.5

4. Which function has the greatest slope?

ANSWER:

(c)

5. Which function has the greatest vertical intercept?

ANSWER:

(a)

6. Which functions have a negative slope?

ANSWER:

None

7. Which function has a zero slope?

ANSWER:

(a)

8. It is possible for the graph of a linear function to intersect

(i)   The x-axis only   (ii)   The y-axis only   (iii)   Both the x- and y-axes (iv)   No axis

(a) (i)
(b) (ii)
(c) (i) and (ii)
(d) (iii)
(e) (iv)

ANSWER:

(c)

COMMENT:

Ask students if it possible to draw a line that intersects the x-axis only and why this is not listed as one of the possibilities.

Use Figure 1.6 to answer Problems 9–10.

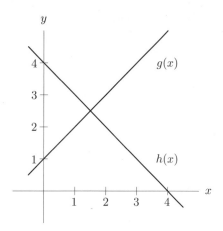

Figure 1.6

9. If $f(x) = h(x) - g(x)$ then the x-intercept of f is at

(a)   0   (b)   1   (c)   2   (d)   3   (e)   None of the previous

ANSWER:

(e)

COMMENT:

The x-axis intercept of f is when $f(x) = 0$, this occurs at the point of intersection. The x value at the point of intersection is seen as between 1 and 2, but not either value.

10. If $f(x) = h(x) - g(x)$ then the y-intercept of f is at

(a)   0   (b)   1   (c)   2   (d)   3   (e)   None of the previous

ANSWER:

(d)

COMMENT:

Ask what type of function $h(x) - g(x)$ is.

ConcepTests and Answers and Comments for Section 1.4

1. Figure 1.7 shows the graph of $f(x) = b + ax$. Determine a and b.

   (a) $a = 1$ and $b = 2$
   (b) $a = 2$ and $b = 1$
   (c) $a = 2$ and $b = 2$
   (d) $a = 0$ and $b = 2$
   (e) None of the above

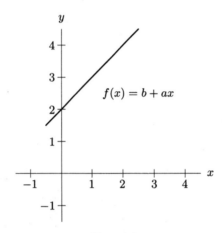

Figure 1.7

ANSWER:

(a)

COMMENT:

As a follow up question ask students to draw lines corresponding to the other answers.

Insert each point in Problems 2–7 into Table 1.2. Then determine if the three data points in the table describe a (a) linear function (b) non-linear function, (c) non-function.

Table 1.2

x	1	2	
y	2	1	

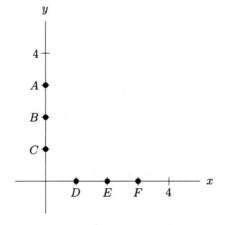

Figure 1.8

2. A

    ANSWER:

    (a)

3. B

    ANSWER:

    (b)

4. C

    ANSWER:

    (b)

5. D

    ANSWER:

    (c)

6. E

    ANSWER:

    (c)

7. F

    ANSWER:

    (a)

8. In the equation $3x + 2y = 9$, if x increases by 2 then y

    (a) Increases by 2
    (b) Decreases by 6
    (c) Increases by 3
    (d) Increases by 6
    (e) Decreases by 3

    ANSWER:

    (e)

    COMMENT:

    This is a good problem to review the concept of an equation.

For Problems 9–15, identify whether the equations is

  (a) Slope-intercept form

  (b) Point-slope form

  (c) Standard form

  (d) None of the above

9. $x + 2y - 5 = 0$

    ANSWER:

    (c)

10. $2y + x = 7$

    ANSWER:

    (d)

11. $y = \dfrac{x - 2}{5}$

    ANSWER:

    (b)

    COMMENT:

    It could be argued that since it is not written exactly in the form $y - 0 = \frac{1}{5}(x - 2)$ that it is (d) None of the above.

12. $2y = x + 5$

    ANSWER:

    (d)

    COMMENT:

    This is close to the slope-intercept form, but note that the slope and intercept are not explicit until both sides have been divided by 2.

13. $y = -5(x - 2)$

    ANSWER:

    (b)

    COMMENT:

    What is the point that is used in this equation?

14. $y = 10 - 5x$

    ANSWER:

    (a)

    COMMENT:

    Note the relationship to Problem 13

15. $y = x$

    ANSWER:

    (a) or (b)

    COMMENT:

    Ask how to convert to standard form.

16. If $x + 2y - 5 = 0$ then an equation of the same line is

    (a) $y = -\frac{1}{2}x + \frac{5}{2}$

    (b) $3x + 6y = 15$

    (c) $y - 0 = -\frac{1}{2}(x - 5)$

    (d) All of the above

    (e) None of the above

       ANSWER:

       (d)

       COMMENT:

       Ask students how many different equations describe the same line.

ConcepTests and Answers and Comments for Section 1.5 ———————

1. The graph of a linear function can be

    (a) Diagonal

    (b) Horizontal

    (c) Vertical

    (d) Diagonal and horizontal

    (e) All of the above

       ANSWER:

       Only (d).

       COMMENT:

       A vertical line is not the graph of a function.

2. The weekly cost function for a woodworker making tables is $T = 20n + 50$, where n is the number of tables made. In the equation, which of the following includes the rent on the workshop?

    (a) T only

    (b) 20

    (c) n

    (d) T and 50

    ii 20 and 50

       ANSWER:

       Only (d).

       COMMENT:

       Ask students what are the practical meanings of 20, 50, T and n in this problem are.

3. The weekly cost function for a woodworker making tables is $T = 20n + 50$. Which of the following includes the cost of lumber for a table?

   (a) T only
   (b) 20 and T
   (c) n and T
   (d) 50

   ANSWER:
   Only (b).

4. Which of the following equations is best for a phone card user to calculate the balance on a $20 phone card with a rate of 5 cents a minute.

   (a) $B = 5m + 20$
   (b) $B = 20 - 0.05m$
   (c) $B = \frac{1}{20}m - 20$
   (d) $B = 20 - 5m$

   ANSWER:
   (b)
   COMMENT:
   Ask students to come up with situations that could be described by the other answers.

5. Two lines can intersect in how many points?

   (a) None
   (b) One
   (c) Two
   (d) Infinite
   (e) All but (c)

   ANSWER:
   (e)
   COMMENT:
   Ask students to draw each situation.

6. Using the data in Table 1.3 identify pairs of linear functions that are perpendicular.

   Table 1.3

x	0	1	2	3
$r(x)$	1/2	5/2	9/2	13/2
$s(x)$	1	-1	-3	-5
$t(x)$	1	1/2	0	$-1/2$
$u(x)$	5	7	9	11

   (a) $r(x)$ and $u(x)$
   (b) $r(x)$ and $t(x)$ only
   (c) $t(x)$ and $u(x)$ only
   (d) $r(x)$ and $t(x)$ and $t(x)$ and $u(x)$
   (e) $s(x)$ and $s(x)$

   ANSWER:
   (d)

ConcepTests and Answers and Comments for Section 1.6 ———————————

1. In Figure 1.9, which of the lines is the least square regression line for the four points?

   (a) A
   (b) B
   (c) C
   (d) D

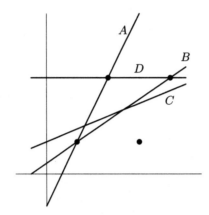

Figure 1.9

ANSWER:

(c)

2. In Figure 1.10, the correlation coefficients in increasing order of r are

    (a) (III), (II), (I), (IV)

    (b) (IV), (II), (I), (III)

    (c) (I), (IV), (II), (III)

    (d) None of the above

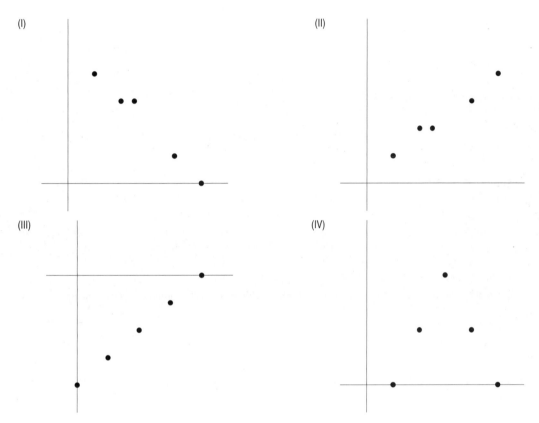

Figure 1.10

ANSWER:

(c)

3. For each time, given in (i) to (iv), decide whether using the regression equation, derived from Table 1.4 to predict weight, is an (a) interpolation or an (b) extrapolation.

    (i) 0   (ii) 6   (iii) 18   (iv) 24

Table 1.4

Time (hours)	0	5	10	15
Weight (gm)	0.5	7.5	13.7	20.2

ANSWER:

  (i) (a)

 (ii) (a)

(iii) (b)

(iv) (b)

COMMENT:

Note that for (i) the data gives the exact answer while the regression line will give an approximate answer.

4. If a value of a regression function is not an interpolation, it is an extrapolation.

   (a) True
   (b) False

      ANSWER:
      (a) True

5. A regression equation can only be found for positive data.

   (a) True
   (b) False

      ANSWER:
      (b) False

6. Correlation and causation are the same.

   (a) True
   (b) False

      ANSWER:
      (b) False

7. Scatter plots are graphs of functions.

   (a) True
   (b) False

      ANSWER:
      (b) False
      COMMENT:
      Some scatter plots are graphs of functions, but not all are.

8. A linear regression equation can only be calculated from a table whose data points describe a function.

   (a) True
   (b) False

      ANSWER:
      (b) False

9. The graph of a linear regression equation must pass through at least one of the data points used to calculate it.

   (a) True
   (b) False

      ANSWER:
      (b) False

Chapter Two

ConcepTests and Answers and Comments for Section 2.1

1. If $r(t) = at^2 + 4$ find $r(2)$.

   (a) $2t^2 + 4$ (b) $4a + 4$ (c) $6a$ (d) $8a$

   ANSWER:
   (b)
   COMMENT:
   For many students function notation is not easy to understand. Is is worth checking if they understand it by asking for specific inputs or outputs.

2. If $m(t) = 3t - 5$ then $m(0) = ?$

   (a) -5 (b) -2 (c) 0 (d) 3

   ANSWER:
   (a)
   COMMENT:
   For many students function notation is not easy to understand. Is is worth checking if they understand it by asking for specific inputs or outputs.

3. If $m(t) = 3t - 5$ and $m(t) = 0$ then

   (a) $t = -3/5$ (b) $t = 3/5$ (c) $t = -5/3$ (d) $t = 5/3$

   ANSWER:
   (d)
   COMMENT:
   For many students function notation is not easy to understand. Is is worth checking if they understand it by asking for specific inputs or outputs.

4. Match the expressions to the lettered points in Figure 2.1.

   (a) $f(x_1)$
   (b) $f(x_2)$
   (c) $(x_1, f(x_1))$
   (d) $(x_2, f(x_2))$
   (e) $(x_2, f(x_1))$
   (f) $(x_1, f(x_2))$

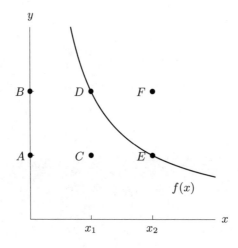

Figure 2.1

   ANSWER:
   (a) B (b) A (c) D (d) E (e) F (f) C

5. If $y = f(x)$, then $f(6)$ is a point on the graph of the function.

   (a) True
   (b) False

   ANSWER:
   (a) False. The point on the graph is $(6, f(6))$.

6. The expression $f(6)$ is read f of 6.

   (a) True
   (b) False

   ANSWER:
   (a) True

7. If $y = h(x)$ then $h(b) - h(a)$ is the component of the slope referred to as the run.

   (a) True
   (b) False

   ANSWER:
   (b) False

8. In the expression $h(b)$, the input is b and the output is h.

   (a) True
   (b) False

   ANSWER:
   (a) False. The output is $h(b)$.

9. For each part (i) - (vi) say whether the expression can be evaluated from Table 2.1, where $T = f(h)$. For part (v) and (vi) say whether the equation can be solved for h. Use (a) yes and (b) no.

   (i) $f(0)$
   (ii) $f(70)$
   (iii) $h(24)$
   (iv) $T(24)$
   (v) $f(h) = 0$
   (vi) $f(h) = 70$

Table 2.1

Hours, h	0	24	48	72
Temperature, T	75	65	85	70

ANSWER:

   (i)   a
   (ii) b
   (iii) b
   (iv) b
   (v)   b
   (vi) a

   COMMENT:
   In part (iv) discuss that T is the dependent variable and not the function name, thus $T(24)$ is not a correct functional notation for $f(24)$.

10. Figure 2.2 shows the graph of the velocity of a cyclist traveling due east from home. For parts (i) - (v), say whether the statement is (a) true or (b) false.

    (i) If $v(t) = 0$, the cyclist is at home.
    (ii) $v(0) = v(30)$.
    (iii) Since $v(15) > v(30)$, the cyclist is further from home at $t = 15$ than $t = 30$.
    (iv) At some time, t, the velocity is 10.
    (v) The cyclist might have stopped for lunch after riding 30 minutes.

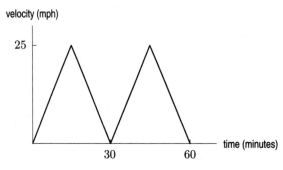

Figure 2.2

ANSWER:

    (i) (b) False
    (ii) (a) True
    (iii) (b) False
    (iv) (a) True
    (v) (b) False

COMMENT:

Ask students to come up with a story for describing the trip of the bicyclist. Students often confuse the path (trajectory) of the bike with the shape of the graph.

11. The US Postal Service has a formula to calculate the price to send a standard envelope within the US. What is the input of this function?

    (a) Dollars
    (b) Destination
    (c) Weight
    (d) Dimensions

ANSWER:
    (c)
COMMENT:
Discuss the output.

12. There is a formula to convert a temperature in Fahrenheit to a temperature in Celsius. What is the input?

    (a) Degrees in Fahrenheit
    (b) Degrees in Celsius
    (c) Both
    (d) Neither

ANSWER:
    (a)

13. From the graph of $s(x)$ in Figure 2.3 determine whether each expression is (a) positive, (b) negative or (c) zero.

  (i) $s(2) - s(1)$
  (ii) $s(3) - s(1)$
  (iii) $s(4) - s(3)$
  (iv) $s(1) - s(4)$

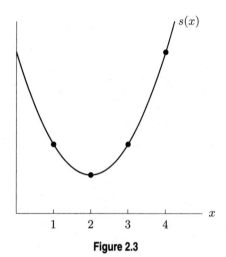

Figure 2.3

ANSWER:

  (i) (b) Negative
  (ii) (c) Zero
  (iii) (a) Positive
  (iv) (b) Negative

COMMENT:
Ask students to draw in the quantities from the problem.

ConcepTests and Answers and Comments for Section 2.2

1. On the graph of a function, the domain is read from the horizontal axis.

    (a) True
    (b) False

    ANSWER:
    (a) True

2. On the graph of a function, the range is read from the positive values on the vertical axis.

    (a) True
    (b) False

    ANSWER:
    (a) False

3. Any value shown on the horizontal axis is in the domain of a function.

    (a) True
    (b) False

    ANSWER:
    (a) False

4. Only values shown on the horizontal axis are in the domain of a function.

   (a) True
   (b) False

   > ANSWER:
   > (a) False
   > COMMENT:
   > A domain of all real numbers is never fully shown.

5. The range of a function can be the same as its domain.

   (a) True
   (b) False

   > ANSWER:
   > (a) True

6. The domain of a linear function is all real numbers.

   (a) True
   (b) False

   > ANSWER:
   > (a) True

7. The range of a linear function is all real numbers.

   (a) True
   (b) False

   > ANSWER:
   > (a) False, remember constant functions.

8. The domain of the function $f(x) = \dfrac{1}{x-2}$ does not include which of the following values.

   (a) -2
   (b) 0
   (c) 2
   (d) π

   > ANSWER:
   > (c)

9. A pot of boiling water is removed from the stove. The temperature of the water, in degrees Fahrenheit, is a function of time, t, in minutes since being taken off the stove. An experiment uses this function until the water temperature falls close to room temperature. What is the most reasonable domain for this experiment?

   (a) $0 \le t \le 1$
   (b) $0 \le t \le 212$
   (c) $0 \le t \le 60$
   (d) $72 \le t \le 212$
   (e) All reals

   > ANSWER:
   > (c)
   > COMMENT:
   > What would be a reasonable range for this function?

10. A pot of boiling water is removed from the stove. The water temperature, C in degrees Celsius, is a function of time, t, in minutes since being taken off the stove. What is the most reasonable range for this function?

    (a) $-10 \le C \le 10$
    (b) $0 \le C \le 212$
    (c) $0 \le C \le 60$
    (d) $20 \le C \le 100$
    (e) All reals

    > ANSWER:
    > (d)
    > COMMENT:
    > What would be a reasonable domain for this function?

11. Find the domain of the function $h(t) = \sqrt{t - 11}$.

   (a) All Reals   (b) All non-negative reals   (c) $0 \leq t < \infty$

   (d) $11 < t < \infty$ (e) $11 \leq t < \infty$

   ANSWER:

   (e)

   COMMENT:

   Remind students that the $t = 11$ is in the domain because the square root of zero is zero.

12. Which of the following represents the domain of the function graphed in Figure 2.4

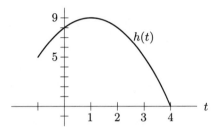

Figure 2.4

   (a) $0 \leq t \leq 4$
   (b) All real numbers
   (c) $0 \leq t \leq 9$
   (d) $-1 \leq t \leq 4$

   ANSWER:
   (d)

ConcepTests and Answers and Comments for Section 2.3

Do the pairs of functions in Problems 1–5 have the same graph? Use (a) yes (b) no.

1. $f(x) = \begin{cases} x^2 - 2 & \text{if} \quad x \geq 0 \\ x & \text{if} \quad x < 0 \end{cases}$ and $g(x) = \begin{cases} x^2 - 2 & \text{if} \quad x > 0 \\ x & \text{if} \quad x \leq 0 \end{cases}$

   ANSWER:
   (b) No

2. $p(x) = \begin{cases} x^2 - 2 & \text{if} \quad x \geq 2 \\ x & \text{if} \quad x < 2 \end{cases}$ and $q(x) = \begin{cases} x^2 - 2 & \text{if} \quad x > 2 \\ x & \text{if} \quad x \leq 2 \end{cases}$

   ANSWER:
   (a) Yes

3. $h(x) = \begin{cases} x^2 - 2 & \text{if} \quad x \geq 0 \\ x & \text{if} \quad x < 0 \end{cases}$ and $k(x) = \begin{cases} x & \text{if} \quad x < 0 \\ x^2 - 2 & \text{if} \quad x \geq 0 \end{cases}$

   ANSWER:
   (a) Yes

4. $a(x) = |x|$ and $b(x) = \begin{cases} x & \text{if} \quad x > 0 \\ -x & \text{if} \quad x \leq 0 \end{cases}$

   ANSWER:
   (a) Yes

5. $r(x) = \begin{cases} x^2 - 2 & \text{if} \quad x > 0 \\ x & \text{if} \quad x \leq 0 \end{cases}$ and $s(x) = \begin{cases} x^2 - 2 & \text{if} \quad 0 < x < \infty \\ x & \text{if} \quad 0 > x > -\infty \end{cases}$

   ANSWER:
   (b) No

6. Table 2.2 represents data from which of the following functions?

Table 2.2

x	0	1	2	3
y	0	1	4	9

(a) $f(x) = \begin{cases} x^2 & \text{if} & x > 0 \\ x & \text{if} & x \le 0 \end{cases}$

(b) $g(x) = \begin{cases} x^2 & \text{if} & x > 1 \\ x & \text{if} & x \le 1 \end{cases}$

(c) $h(x) = \begin{cases} x^2 & \text{if} & x \ge 1 \\ -x & \text{if} & x < 1 \end{cases}$

(d) All of the above

(e) None of the above

ANSWER:

(d)

7. Which of the following is all real numbers for the absolute value function?

(a) Both the domain and range

(b) Domain only

(c) Range only

(d) Neither the domain or range

ANSWER:

(b)

Use Figure 2.5 to answer (a) true or (b) false for each equation given in Problems 8–12.

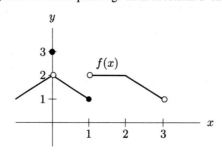

Figure 2.5

8. $f(0) = 3$

ANSWER:

(a) True

9. $f(1/2) = 3/2$

ANSWER:

(a) True

10. $f(1) = 2$

ANSWER:

(b) False

11. $f(2) = 2$

ANSWER:

(a) True

12. $f(3) = 1$

ANSWER:

(b) False

ConcepTests and Answers and Comments for Section 2.4 ━━━━━━━━━

Decide whether each statement in Problems 1–6 is (a) true or (b) false. Assume a is in the domain of an invertible function f.

1. The graph of f contains $(a, f(a))$.
   ANSWER:
   (a) True

2. The graph of f^{-1} contains $(f(a), a)$.
   ANSWER:
   (a) True

3. The graph of f contains $(f^{-1}(a), a)$.
   ANSWER:
   (b) False. We cannot assume that a is in the domain of f^{-1}.

4. The graph of f^{-1} contains (a, a^{-1}).
   ANSWER:
   (b) False

5. If $f(a) = b$ then $f^{-1}(b) = a$.
   ANSWER:
   (a) True

6. The domain of f^{-1} is the range of f.
   ANSWER:
   (a) True

7. There is a function f such that $f(x) = f^{-1}(x)$ for all x.

   (a) True
   (b) False

   ANSWER:
   (a) True, $f(x) = x$.

8. The absolute value function has an inverse function.

   (a) True
   (b) False

   ANSWER:
   (b) False

9. Which of the following tables could be data from an invertible function?

   (a)

t	0	24	48	72
$h(t)$	75	75	75	75

   (b)

t	4	3	2	1
$r(t)$	5	10	15	20

   (c)

t	0	2	4	6
$s(t)$	25	0	15	−10

   (d)

t	0	2	4	6
$m(t)$	−5	−10	−15	−20

   ANSWER:
   All but (a).
   COMMENT:
   If no one answers (c) to start a discussion then ask what would happen if you knew that the graph of $s(t)$ was continuous curve.

Use Figure 2.6 to decide whether each equation in Problems 10–16 is (a) true or (b) false.

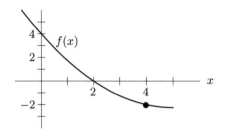

Figure 2.6

10. $f(0) = f^{-1}(-2)$
    ANSWER:
    (a) True

11. $f^{-1}(0) = 4$
    ANSWER:
    (b) False

12. $f^{-1}(2) = 0$
    ANSWER:
    (b) False

13. $f^{-1}(4) = 0$
    ANSWER:
    (a) True

14. $f^{-1}(4) = -2$
    ANSWER:
    (b) False

15. $f^{-1}(3) = -3$
    ANSWER:
    (b) False

16. $f(f^{-1}(3)) = 3$
    ANSWER:
    (a) True

ConcepTests and Answers and Comments for Section 2.5

Are the following statements in Problems 1–10 (a) true or (b) false?

1. Knowing a finite number of points on a graph is never enough to determine whether the graph is concave up or concave down.
    ANSWER:
    (a) True

2. Knowing a finite number of points on a graph can be enough to determine whether the graph is not concave up.
    ANSWER:
    (a) True

3. All functions have graphs that are either concave up or concave down.
    ANSWER:
    (b) False

4. A circle is the graph of a function that is both concave up and concave down.
    ANSWER:
    (b) False, a circle is not the graph of a function.

5. The graph of the absolute value function is concave up.
   ANSWER:
   (b) False

6. The graph of the intensity of the power light on a stereo amplifier, as a function of time since turning off, is concave down.
   ANSWER:
   (b) False, it fades quickly and has a residual glow. The graph decreases but is concave up.

7. The graph of the temperature of a cooling cup of coffee, as a function of time, is concave up.
   ANSWER:
   (a) True

8. The graph of the distance between the station and a departing train, as a function of time, is concave up.
   ANSWER:
   (a) True

9. The graph of the distance between the station and an arriving train, as a function of time, is concave up.
   ANSWER:
   (a) True

10. The graph of a linear function is both concave up and concave down.
    ANSWER:
    (b) False

11. Which of the following tables could be data from a function whose graph is concave up?

    I

t	0	1	2	3
$h(t)$	5	10	15	20

    II

t	0	1	2	3
$r(t)$	−30	−25	−15	0

    III

t	0	2	4	6
$s(t)$	0	15	25	30

    IV

t	0	2	4	6
$m(t)$	5	0	0	5

    (a) I only
    (b) II only
    (c) IV only
    (d) I and III
    (e) II and IV

    ANSWER:
    (b) and (d)
    COMMENT:
    Ask about concavity of the graph for the other two tables. A graph for table (c) could be concave down.

Use Figure 2.7 on the interval given in Problem 12–17 to state whether the graph on that interval is:

(a) Concave up

(b) Concave down

(c) Neither concave up nor concave down

(d) Parts concave up, parts concave down

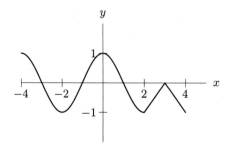

Figure 2.7

12. $0 < x < 1$
    ANSWER:
    (b)

13. $-3 < x < 0$
    ANSWER:
    (d)

14. $-1 < x < 1$
    ANSWER:
    (b)

15. $0 < x < 2$
    ANSWER:
    (d)

16. $2 < x < 4$
    ANSWER:
    (c)

17. $1 < x < 2$
    ANSWER:
    (a)

ConcepTests and Answers and Comments for Section 2.6

1. Which one of the following equations has the graph shown in Figure 2.8?

    (a) $y = 1 - x^2$
    (b) $y = (x - 1)(x - 7)$
    (c) $y = (x + 1)(x + 7)$
    (d) $y = (x - 1)(x + 7)$
    (e) $y = x^2 + x + 1$

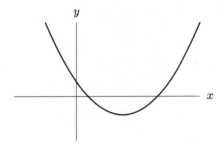

Figure 2.8

ANSWER:
(b)
COMMENT:
Ask students to label the x-intercepts.

2. Which one of the following equations has the graph shown in Figure 2.9.

    (a) $y = x^2 - 1$
    (b) $y = x^2 - 2x + 1$
    (c) $y = x - x^2$
    (d) $y = x^2 + x$
    (e) $y = x^2 - 2x$

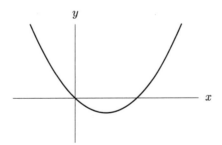

Figure 2.9

ANSWER:
(e)
COMMENT:
Discuss how to eliminate wrong answers.

3. Which one of the following equations has the graph shown in Figure 2.10.

   (a) $y = x^2 - 1$
   (b) $y = x^2 - 2x + 1$
   (c) $y = 1 + x^2$
   (d) $y = x^2 + x$
   (e) $y = x^2 - x$

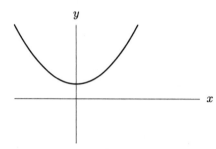

Figure 2.10

ANSWER:
(c)
COMMENT:
This problem is a good preview of shifting graphs of functions.

4. Which one of the following equations has the graph shown in Figure 2.11.

   (a) $y = 1 - x^2$
   (b) $y = -x^2 - 2x + 1$
   (c) $y = -(x - 1)(x + 2)$
   (d) $y = (x + 1)(x - 2)$
   (e) $y = (2 - x)(x + 1)$

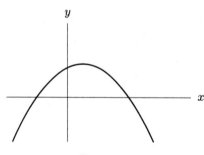

Figure 2.11

ANSWER:
(e)
COMMENT:
Ask students how graphs of (c), (d), and (e) are different.

5. Which one of the following equations has the graph shown in Figure 2.12.
   - (a) $y = (x+1)^2$
   - (b) $y = x^2 - 2x + 1$
   - (c) $y = x - x^2$
   - (d) $y = (x+2)(x+2)$
   - (e) $y = (x-2)(x+2)$

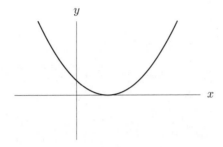

Figure 2.12

ANSWER:

(b)

COMMENT:

This is an example of a polynomial that has a double root, which on the graph manifests itself in the fact that the graph touches the x-axis but does not cross.

Are the statements in Problems 6–16 (a) true or (b) false?

6. All quadratic equations can be expressed in factored form as $y = a(x - r_1)(x - r_2)$.
   ANSWER:
   (b) False
   COMMENT:
   This assumes r_1 and r_2 are real numbers. Avoid a discussion of complex numbers at this point.

7. The quadratic equation $y = x(x+2)$ is expressed in factored form.
   ANSWER:
   (a) True
   COMMENT:
   Purists might demand $y = (x-0)(x+2)$

8. The zeros of a quadratic function are easily determined from its factored form equation.
   ANSWER:
   (a) True

9. A factored form of $y = x^2$ is $y = (x-0)(x-0)$.
   ANSWER:
   (a) True

10. In the general form of a quadratic equation, $y = ax^2 + bx + c$, the a value determines whether the graph of the function will be concave up or concave down.
    ANSWER:
    (a) True

11. The graph of a quadratic function is either concave up or concave down.
    ANSWER:
    (a) True

12. If the average rate of change of a quadratic function changes from negative to positive as x increases then the graph is concave down.
    ANSWER:
    (b) False

13. The domain of a quadratic function is all reals.
    ANSWER:
    (a) True

14. The range of a quadratic function is all reals.
>   ANSWER:
>   (b) False

15. The graph of any quadratic function has at least one x-intercept.
>   ANSWER:
>   (b) False
>   COMMENT:
>   Ask students to give an example of each, 0, 1, or 2 x-intercepts.

16. The graph of any quadratic function has at least one y-intercept.
>   ANSWER:
>   (a) True
>   COMMENT:
>   Can it have more than one?

Chapter Three

ConcepTests and Answers and Comments for Section 3.1 ————

1. For $P = f(t) = ab^t$, with $b > 0$, then a and b are

   (a) Independent variables
   (b) Dependent variables
   (c) Known constants
   (d) Parameters
   (e) Function names

   ANSWER:
   (d)
   COMMENT:
   Ask students to give examples of the other terms in the context of this problem.

2. For $P = f(t) = ab^t$, with $b > 0$, then f is the

   (a) Independent variable
   (b) Dependent variable
   (c) Known constant
   (d) Parameter
   (e) Function name

   ANSWER:
   (e)

3. For $P = f(t) = ab^t$, with $b > 0$ then t is

   (a) Independent variable
   (b) Dependent variable
   (c) Known constant
   (d) Parameter
   (e) Function name

   ANSWER:
   (a)

4. For $P = f(t) = b^t$, with $b > 0$, then P is

   (a) Independent variable
   (b) Dependent variable
   (c) Known constant
   (d) Parameter
   (e) Function name

   ANSWER:
   (b)

5. For exponential growth whose percent rate of change is 3%, what is the growth factor?

   (a) 3.0
   (b) 0.03
   (c) 1.03
   (d) 0.3
   (e) 0.97

   ANSWER:
   (c)

6. For exponential growth whose percent rate of change is -3%, what is the growth factor?

   (a) -3.0
   (b) -0.03
   (c) -1.03
   (d) -0.3
   (e) 0.97

   ANSWER:
   (e)

7. If an exponential function has a growth factor of 3/2, what is the percent rate of change?

   (a) 2.5%
   (b) 3/2%
   (c) 1/2%
   (d) 50%
   (e) 150%

   ANSWER:
   (d)

8. If an exponential function has a growth factor of 0.9, what is the percent rate of change?

   (a) -10%
   (b) -90%
   (c) 10%
   (d) 90%
   (e) None of the above

   ANSWER:
   (a)

9. Match the equation to the description
   A) $P = 1000(1.06)^t$ B) $P = 2(0.94)^t$ C) $P = 0.5(106)^t$ D) $P = (0.06)^t$

   (a) Astronomical growth
   (b) A 6% increasing growth rate
   (c) A 6% decay rate
   (d) Quickly approaches zero

   ANSWER:
   (a) C   (b) A   (c) B   (d) D

ConcepTests and Answers and Comments for Section 3.2

1. The points $(0, 5)$ and $(1, 10)$ are data points for which of the following functions?

   (a) $f(t) = 2 \cdot 5^t$
   (b) $f(t) = 5 \cdot 2^t$
   (c) $f(t) = 5 \cdot 10^t$
   (d) $f(t) = 10 \cdot (\frac{1}{2})^t$

   ANSWER:
   (b)

2. Which of the following are exponential equations?

   (a) $y = 7.5 + 2x$
   (b) $y = 7.5 \cdot 2^x$
   (c) $y = 7.5 \cdot x^2$
   (d) $y = 7.5 \cdot \frac{x}{2}$

   ANSWER:
   Only (b)
   COMMENT:
   Ask the students what the other equations are (linear, quadratic).

Answer Problems 3–6 using Tables (a) and (b).

(a)

t	0	1	2	3
$f(t)$	2	4	8	16

(b)

t	0	1	2	3
$h(t)$	100	150	200	250

3. Which table shows a linear function?

    ANSWER:

    (b)

4. Which table shows an exponential function?

    ANSWER:

    (a)

5. Which table shows the greatest function values?

    ANSWER:

    (b)

6. For large values of t, like $t > 1000$, which table would show the greatest function values?

    ANSWER:

    (a)

    COMMENT:

    This problem points out again that in the long run, exponential functions dominate linear functions.

Answer Problems 7–11 using Tables (a) - (d).

(a)

t	0	1	2	3	4
$f(t)$	1	2	4	8	16

(c)

t	0	1	2	3	4
$h(t)$	0.1	0.2	0.4	0.8	1.6

(b)

t	0	1	2	3	4
$g(t)$	−1	−2	−4	−8	−16

(d)

t	0	1	2	3	4
$s(t)$	10	20	40	80	160

7. All four functions are exponential.

    (a) True   (b) False

    ANSWER:

    (a)

8. All four functions have the same base.

    (a) True   (b) False

    ANSWER:

    (a)

9. All four functions are have the same value for a, when in the form $f(t) = a \cdot b^t$.

    (a) True   (b) False

    ANSWER:

    (b)

10. In Table (c), with $h(t) = a \cdot b^t$, the value for a is?

    (a) 1.0

    (b) 0.01

    (c) 1/10

    (d) 2

    (e) None of the above

    ANSWER:

    (c)

11. The percent growth rate in each of the functions is?

    (a) Different in each case   (b) 1   (c) 50%   (d) 100%   (e) 200%

    ANSWER:

    (d), Since $b = 2$, then $r = 1$, and the rate is 100%.

Answer Problems 12–14 using Table 3.1.

Table 3.1

t	1	2	3	4
$f(t)$	8	4	2	1

12. The percent growth rate of the function is?
    (a) 100%   (b) 50%   (c) −50%   (d) −100%   (e) −200%

    ANSWER:
    (c)

13. The base of the exponential function is?
    (a) 1/2   (b) 2   (c) −1/2   (d) −2   (e) None of the previous

    ANSWER:
    (a)

14. If $f(t) = a \cdot b^t$, then $a =$?
    (a) 1   (b) 2   (c) 4   (d) 8   (e) None of the previous

    ANSWER:
    (e) $a = 16$.

ConcepTests and Answers and Comments for Section 3.3

Identify the statements in Problems 1–7 as (a) true or (b) false.

1. Infinity is a large real number.
    ANSWER:
    (b) False

2. The notation $f(x) \to k$ means $f(x)$ will eventually be equal to k.
    ANSWER:
    (b) False

3. For exponential growth, the percent rate of change can easily be determined from the growth rate.
    ANSWER:
    (a) True

4. It is possible that $f(x) \to -\infty$ as $x \to \infty$.
    ANSWER:
    (a) True
    COMMENT:
    Have students give examples.

5. Interpolation and extrapolation can be used with an exponential regression equation.
    ANSWER:
    (a) True

6. An exponential regression equation can not be calculated on linear data.
    ANSWER:
    (b) False

7. A population, in millions, is modeled by $f(t) = 10 \cdot (1.03)^t$, where t is years since 2000. To graphically determine when the population reaches 15 million, look at the intersection of the graph of $y = f(t)$ and the line $t = 15$.
    ANSWER:
    (b) False, graph the horizontal line $y = 15$.
    COMMENT:
    Ask students what $t = 15$ helps you find.

8. An exponential function has a vertical asymptote.

   (a) Sometimes   (b) Always   (c) Never

   ANSWER:
   (c)

9. An exponential function has a horizontal asymptote.

   (a) Sometimes   (b) Always   (c) Never

   ANSWER:
   (b)

The four functions shown in Figure 3.1 are exponential and each is defined by an equation the form $f(t) = ab^t$. Use this information to answer Problems 10–13.

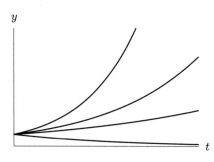

Figure 3.1

10. The number of functions shown with $b < 1$ is

    (a) Zero   (b) 1   (c) 2   (d) 3   (e) 4

    ANSWER:
    (b)

11. Which of the following statement(s) is/are true? The values of a in the functions shown are

    (I) All the same
    (II) Positive
    (III) Negative
    (IV) Zero
    (V) None of the above

    (a) I only
    (b) IV only
    (c) I and III
    (d) I and II
    (e) V

    ANSWER:
    (d)

12. The values of b in the functions shown are

    (a) All the same
    (b) All positive
    (c) All negative
    (d) Three positive, one negative

    ANSWER:
    (b)

13. The number of functions shown having $y = 0$ as a horizontal asymptote is

    (a) Zero   (b) 1   (c) 2   (d) 3   (e) 4

    ANSWER:
    (e)

The three functions shown in Figure 3.2 are exponential and each is defined by an equation of the form $f(t) = ab^t$. Use this information to answer Problems 14–17 .

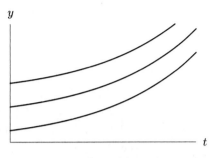

Figure 3.2

14. The number of functions with $b < 1$ is

    (a) Zero
    (b) 1
    (c) 2
    (d) 3

    ANSWER:
    (a)

15. The parameters a for the functions shown are

    (a) All the same
    (b) All positive
    (c) All negative
    (d) None of the above

    ANSWER:
    (b)

16. The parameters b are

    (a) All the same
    (b) All positive
    (c) All negative
    (d) None of the above

    ANSWER:
    (b)
    COMMENT:
    To determines that all b values are not the same (without seeing the function formula) notice that the distances between the graphs remains constant as t increases. This means that $a_1 b^t - a_2 b^t = (a_1 - a_2)b^t$ is constant. But b^t is not constant, since it is clear from the graph that $b > 1$. Thus, the b parameters are not all the same, since, if they were, then the distance between graphs for given t values would increase as t increases.

17. The number of functions shown which have a horizontal asymptote at $y = 0$ is

    (a) Zero
    (b) 1
    (c) 2
    (d) 3

    ANSWER:
    (d)

ConcepTests and Answers and Comments for Section 3.4 ————————

Identify the statements in Problems 1–4 as (a) true or (b) false.

1. The parameters of $f(t) = ae^{kt}$ are a, e, and k.
   ANSWER:
   (b) False, e is a constant.

2. By looking at the graph of an exponential function, it is not possible to know whether the equation is written in the form $f(t) = ae^{kt}$ or in the form $f(t) = ab^t$.
   ANSWER:
   (a) True
   COMMENT:
   You can always write the equation either way.

3. In the formula $f(t) = ae^{kt}$, if $k > 1$ the function is increasing and if $k < 1$ then the function is decreasing.
   ANSWER:
   (b) False
   COMMENT:
   Note that a could be positive or negative in this problem.

4. If $f(t) = g(t)$ where $f(t) = a_0 b^t$ and $g(t) = a_1 e^{kt}$ then $a_0 = a_1$.
   ANSWER:
   (a) True
   COMMENT:
   We also know that $b = e^k$.

5. If the graph of $y = ab^t$ is a line then

   (a) $a = 1$
   (b) $b = 1$
   (c) $t = 1$
   (d) None of the above

   ANSWER:
   (b)
   COMMENT:
   Note that $b = 0$ is not an option when we talk about exponential functions.

6. Order the numbers e^2, 2^3, 3^2, 5^0 from smallest to largest.

   (a) $3^2, 2^3, e^2, 5^0$
   (b) $5^0, e^2, 2^3, 3^2$
   (c) $5^0, 2^3, e^2, 3^2$
   (d) $5^0, e^2, 3^2, 2^3$

   ANSWER:
   (b)

7. Out of the given options, what is the value of b in Figure 3.3?

   (a) $b = 0$ (b) $b = 1$ (c) $b = 2$ (d) $b = 3$

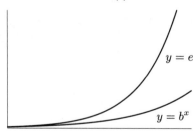

Figure 3.3

   ANSWER:
   (c)

ConcepTests and Answers and Comments for Section 3.5

Identify the statements in Problems 1–3 as (a) true or (b) false.

1. Given a birth present of a $50 savings bond earning 4% compounded quarterly, to find if you will be a millionaire at age 90, enter the following on a calculator: $50 * 1.04^{\wedge}90$.

    ANSWER:

    (b) False

    COMMENT:

    Use $50 * 1.01^{\wedge}(4 * 90)$ to account for quarterly compounding.

2. If $1000 is put into account A at 3% interest, compounded annually, and $10 is put into account B at 3% interest, compounded continuously, then account A will always have a greater value than account B.

    ANSWER:

    (b) False

    COMMENT:

    Eventually account B will overtake account A, but it will take more than 10,000 years.

3. If $1000 is put into account A at 4% interest, compounded annually, and $10 is put into account B at 3% interest, compounded continuously, then account A will always have a greater value than account B.

    ANSWER:

    (a) True

4. One dollar compounded quarterly at 8% annual interest has a balance after t years of

    (a) 1.08^t
    (b) 1.08^{4t}
    (c) 1.02^{4t}
    (d) $(1 + \frac{0.08}{4})^t$
    (e) None of the above

    ANSWER:

    (c)

    COMMENT:

    Ask the students to come up with situations that the other formulas could belong to.

Chapter Four

ConcepTests and Answers and Comments for Section 4.1 ────────

For Problems 1–11, identify the statement as (a) true or (b) false.

1. A logarithmic function and an exponential function can be inverses of one another.
    ANSWER:
    (a) True

2. The domain of a logarithmic function is all real numbers.
    ANSWER:
    (b) False

3. The range of a logarithmic function is all real numbers.
    ANSWER:
    (a) True

4. $\log 10 = 1$
    ANSWER:
    (a) True

5. $\log(x + 2) = \log x + \log 2$
    ANSWER:
    (b) False

6. $\log(2x) = \log x + \log 2$
    ANSWER:
    (a) True

7. $\log(x^2) = 2 + \log x$
    ANSWER:
    (b) False

8. $\log(2^x) = x \cdot \log 2$
    ANSWER:
    (a) True

9. The equation $10^{2.7t} = 15$ can be solved for t by taking the log of both sides and then dividing both sides by 2.7.
    ANSWER:
    (a) True

10. The equation $10^{3.5t} = -105$ can be solved for t by taking the log of both sides and then dividing both sides by 3.5.
    ANSWER:
    (b) False
    COMMENT:
    There is no solution.

11. There is a positive constant k so that $\ln x = k \cdot \log x$ for all x.
    ANSWER:
    (a) True
    COMMENT:
    Note that $k = \ln 10$.

12. Match the curves A, B, C and D in Figure 4.1 with the equations (I) to (IV).

    (I) $y = \ln(x)$
    (II) $y = \ln(x + 2)$
    (III) $y = \ln(2x)$
    (IV) $y = \ln(x/2)$

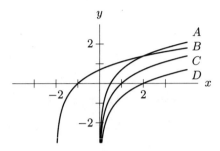

Figure 4.1

    ANSWER:

    (I) C
    (II) B
    (III) A
    (IV) D

13. Without using a graphing utility, identify the function f in Table 4.1 as either (a) $f(x) = \ln(x)$ or (b) $f(x) = \log(x)$.

Table 4.1

x	1	2	3	4
$f(x)$	0	0.6932	1.0986	1.3863

    ANSWER:
    (a) $f(x) = \ln(x)$
    COMMENT:
    Ask the students what values of x give $0 < \log x < 1$ and what values of x give $0 < \ln x < 1$.

ConcepTests and Answers and Comments for Section 4.2

For Problems 1–5, identify the statement as (a) true or (b) false.

1. A one million dollar investment will double in the same time as a ten dollar investment if the interest rates are the same.
    ANSWER:
    (a) True

2. Any exponential equation whose base is e can be equivalently represented by an exponential equation whose base is 10.
    ANSWER:
    (a) True

3. If $f(t) = a_1 b^t = a_2 e^{kt}$, for all t, then $a_1 = a_2$.
    ANSWER:
    (a) True

4. The equation $B = 1000e^{1.06t}$ represents a 6% continuously compounded return on $1000.
    ANSWER:
    (b) False

5. If an equation can be solved using $\log(x)$, it can also be solved by using $\ln(x)$.
    ANSWER:
    (a) True
    COMMENT:
    Ask the students for some reasons why they would use one of the other in a given situation.

6. If an investment has a 6% return, compounded annually, in how many years will it double?

   (a) 3

   (b) $\dfrac{\log(2)}{\log(1.06)}$

   (c) $\dfrac{\log(2)}{\log(0.06)}$

   (d) $\dfrac{\log(1.06)}{\log(2)}$

   (e) None of the above

   ANSWER:

   (b)

7. If a chemical compound decays at a continuous rate of 3% per year, what is its half-life?

   (a) $\dfrac{\ln(0.5)}{\ln(1.03)}$

   (b) $\dfrac{\ln(0.5)}{\ln(0.03)}$

   (c) $\dfrac{-\ln(2)}{\ln(0.03)}$

   (d) $\dfrac{\ln(2)}{0.03}$

   (e) None of the above

   ANSWER:

   (d)

   COMMENT:

   Good time to review $\ln(1/2) = -\ln(2)$.

8. If a quantity is increasing at a continuously compounded rate and a second quantity is decreasing at the same rate, then the doubling time of the first quantity is related to the half-life of the second quantity by which of the following statements

   (a) The doubling time is greater than the half-life
   (b) The half-life is greater than the doubling time
   (c) The half-life is equal to the doubling time
   (d) The relationship varies with the rate
   (e) None of the above

   ANSWER:

   (c)

   COMMENT:

   Get students started by asking them to think about certain situations.

For Problems 9–16, classify the equations using the following statements:

(a) Solvable using an exponential function

(b) Solvable using a logarithmic function

(c) Not solvable using an exponential or logarithmic function, but solutions exist

(d) No real number is a solution

9. $2.5e^{0.03t} = 3.7$

    ANSWER:

    (b)

10. $-2.5e^{0.06t} = 7.5$

    ANSWER:

    (d)

11. $2e^{0.05t} = 5t$

    ANSWER:

    (c)

    COMMENT:

    This can be illustrated by graphing both sides of the equations. Since the graphs intersect, a solution exists.

12. $80(1.03)^t = 40$

    ANSWER:

    (b)

13. $80e^{-0.03t} = 40$

    ANSWER:

    (b)

14. $\ln(2x + 1) = x$

    ANSWER:

    (c)

15. $2\ln(0.5x - 1) = 7.5$

    ANSWER:

    (a)

16. $\log(x) = -7.5$

    ANSWER:

    (a)

ConcepTests and Answers and Comments for Section 4.3 ────────────

For Problems 1–9, identify the statement as (a) true or (b) false.

1. The domain of $\log(x)$ is all real numbers.
     ANSWER:
     (b) False

2. The range of $\log(x)$ is all real numbers.
     ANSWER:
     (a) True

3. There is no solution to the equation $\log(x) = e^x$.
     ANSWER:
     (a) True
     COMMENT:
     Look at the graphs of the two function to see that they do not intersect.

4. The log function has a horizontal and a vertical asymptote.
     ANSWER:
     (b) False

5. The graph of $y = \log(e^x)$ is linear.
     ANSWER:
     (a) True
     COMMENT:
     Since $\log(e^x) = x\log(e)$ and e is a constant, $y = \log(e^x)$ is a linear.

6. If the scale is the same on both axes, the graphs of a function and its inverse are symmetric across the line $y = x$.
     ANSWER:
     (a) True

7. For a specific application of the decibel equation, $D = 10 \cdot \log(I/I_0)$, I_0 is the independent variable.
     ANSWER:
     (b) False
     COMMENT:
     Note that I is the independent variable.

8. The decibel equation, $D = 10 \cdot \log(I/I_0)$, can also be written as $D = 10 \cdot \log(I) - 10 \cdot \log(I_0)$.
     ANSWER:
     (a) True

9. If one sound is 5 decibels and a second sound is 10 decibels, the second sound is twice as loud as the first.
     ANSWER:
     (b) False
     COMMENT:
     Ask the students how much louder the second sound is.

10. If a sound is increased by an order of magnitude then it is how many times louder.
     (a) 10%
     (b) 10
     (c) 2
     (d) 100
     (e) None of the above
        ANSWER:
        (b)

11. If a sound is decreased by an order of two magnitudes then it is how many times softer.
     (a) 2%
     (b) 10
     (c) 2
     (d) 100
     (e) None of the above
        ANSWER:
        (d)

12. Without using a graphing utility, match each pair of equations to a graph in Figure 4.2. All scales are equal.

(I) $y = e^x$ and $y = \log x$

(II) $y = e^x$ and $y = \ln x$

(III) $y = 10^x$ and $y = \log x$

(IV) $y = 10^x$ and $y = \ln x$

(a)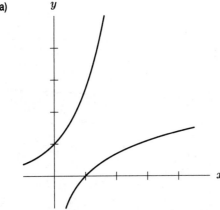

(b)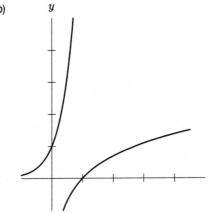

(c)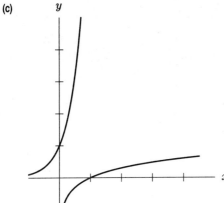

(d)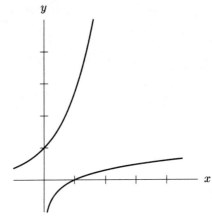

Figure 4.2

ANSWER:

(I) (d)

(II) (a)

(III) (c)

(IV) (b

COMMENT:

Ask about tick mark values in the figures.

ConcepTests and Answers and Comments for Section 4.4 ───────────

For Problems 1–11, identify the statement as (a) true or (b) false.

1. The graph of the log function is concave down.
    ANSWER:
    (a) True

2. The log function has negative values in it range.
    ANSWER:
    (a) True

3. The log function is increasing.
    ANSWER:
    (a) True

4. The function $y = -\ln(x)$ is concave down.
    ANSWER:
    (b) False

5. On a log scaled axis the tick values are always labeled in exponent format.
    ANSWER:
    (b) False

6. Log scaling cannot be used on only one axis.
    ANSWER:
    (b) False

7. The graph of an exponential function appears linear when the vertical axis is changed to log scale.
    ANSWER:
    (a) True

8. Two tick marks at 5 and 10, are the same distance apart as two tick marks at 500 and 1000 on a log scaled axis.
    ANSWER:
    (a) True

9. Two tick marks at 4 and 5, are the same distance apart as two tick marks at 9 and 10 on a log scaled axis.
    ANSWER:
    (b) False

10. Two tick marks at 4.5 and 5.7, are the same distance apart as two tick marks at 9.1 and 10.3 on a linearly scaled axis.
    ANSWER:
    (a) True

11. A log scaled axis has no zero point.
    ANSWER:
    (a) True

For each table in Problems 12–16, identify the best scale choice for graphing the function.

(a) Linear scale on both axes

(b) Log scale on both axes

(c) Log scale on horizontal axes, linear scale on vertical axis

(d) Log scale on vertical axes, linear scale on horizontal axis

12.

Table 4.2

t	1	5	755	837	15000
$f(t)$	−74	−96	−84	−72	−92

ANSWER:
(c)

13.

Table 4.3

t	1600	1700	1800	1900	2000
$g(t)$	$7.5 \cdot 10^6$	$8.4 \cdot 10^6$	$9.5 \cdot 10^6$	$1.3 \cdot 10^7$	$2.5 \cdot 10^7$

ANSWER:
(a), The numbers are large but not varied.

14.

Table 4.4

x	0.0002	0.0012	0.3	0.14	2.2
$h(x)$	$7.5 \cdot 10^{-6}$	$7.5 \cdot 10^{-6}$	$1.5 \cdot 10^{-6}$	$1.1 \cdot 10^6$	$1.5 \cdot 10^7$

ANSWER:
(b)

15.

Table 4.5

x	1	2	3	4	5
$r(x)$	$-7.5 \cdot 10^{-6}$	$-1.5 \cdot 10^{-5}$	$-1.7 \cdot 10^{-6}$	$-1.1 \cdot 10^{-2}$	$-1.5 \cdot 10^{-2}$

ANSWER:
(d)

16.

Table 4.6

x	6	12	3000	5000
$n(x)$	80000	80000	80000	80000

ANSWER:
(c)

Chapter Five

ConcepTests and Answers and Comments for Section 5.1 ─────

1. The functions $f(x)$ and $f(x) + c$ always have the same

   (a) Zeros
   (b) Vertical intercept
   (c) Domain
   (d) Range
   (e) None of the above

   ANSWER:
   (c)
   COMMENT:
   Ask the students how the two functions are related.

2. The function $f(x)$ and $f(x + c)$ always have the same

   (a) Zeros
   (b) Vertical intercept
   (c) Domain
   (d) Range
   (e) None of the above

   ANSWER:
   (d)
   COMMENT:
   Ask the students how the two functions are related.

For Problems 3–11, identify the statement as (a) true or (b) false.

3. The functions $f(x)$ and $f(x + c)$ always have the same zeros.
   ANSWER:
   (b) False
   COMMENT:
   Ask the students how the zeros are related.

4. The graph of $f(x + c)$, with $c > 0$, is the graph of $f(x)$ but shifted to the right by c units.
   ANSWER:
   (b) False
   COMMENT:
   The graph is shifted to the left by c units.

5. The graph of $f(x) + c$, with $c < 0$, is the graph of $f(x)$ but shifted upward by c units.
   ANSWER:
   (b) False
   COMMENT:
   Since $c < 0$ the graph is shifted downward.

6. If $f(x) = e^x$, then $f(x + c) = e^x + c$.
   ANSWER:
   (b) False
   COMMENT:
   Note that c is part of the input. This is a good point to review function notation, input and output.

7. If $f(x) = \ln(x)$, then $f(x + c) = \ln(x) + \ln(c)$.
   ANSWER:
   (b) False

8. If $f(x) = e^x$, then $f(x) + 1 = e^x + 1$.
   ANSWER:
   (a) True

9. If $f(x) = e^x$, then $f(x + 1) = e^{x+1}$.
   ANSWER:
   (a) True

10. Let $g(x) = a + f(x + b)$, where a and b are positive. Then $g(x)$ has the same graph as $f(x)$, but shifted horizontally by a units and vertically by b units.

   ANSWER:

   (a) True

11. The graph of a function first shifted horizontally, then shifted vertically results in the same graph if the order of the shifts is reversed.

   ANSWER:

   (a) True

12. Which of the following functions have the same range as the function $f(t)$?

   (I) $g(t) = f(t + 4)$
   (II) $h(t) = f(t - 3)$
   (III) $r(t) = f(t) + 2$
   (IV) $s(t) = f(t) - 1$

   (a) I and II
   (b) III and IV
   (c) II and IV
   (d) I and III

   ANSWER:

   (a)

13. Which two of the following are always equal?

   (I) $f(2 + 3)$
   (II) $f(2) + 3$
   (III) $f(3 + 2)$
   (IV) $f(3) + 2$
   (V) None

   (a) I and II
   (b) I and III
   (c) I and IV
   (d) III and IV
   (e) V

   ANSWER:

   (b)

ConcepTests and Answers and Comments for Section 5.2 ———

1. If $f(x) = e^{-x}$, match the following functions to the graphs in Figure 5.1

   (I) $f(x)$
   (II) $u(x) = -f(x)$
   (III) $v(x) = f(-x)$
   (IV) $w(x) = -f(-x)$

(a)

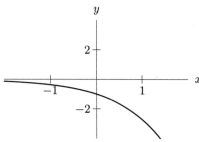

(b)

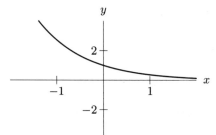

(c)

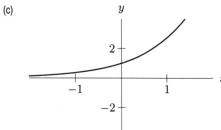

(d)

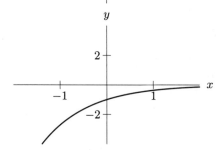

Figure 5.1

ANSWER:

(I) b
(II) d
(III) c
(IV) a

2. If $f(x) = (x+1)^2 - 2$, match the following functions to the graphs in Figure 5.2

   (I) $f(x)$
   (II) $u(x) = -f(x)$
   (III) $v(x) = f(-x)$
   (IV) $w(x) = -f(-x)$

(a)

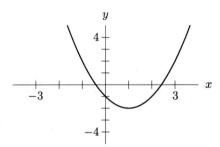

(b)

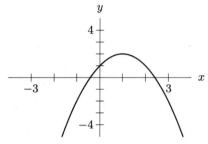

(c)

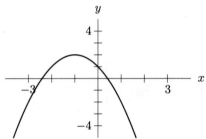

(d)

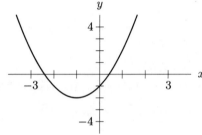

Figure 5.2

   ANSWER:

   (I) d
   (II) c
   (III) a
   (IV) b

3. What is the vertical intercept of $g(t) = f(-t)$, if $f(t) = 2e^{0.07t}$? Answer without using a calculator or computer.

   (a) 2
   (b) -2
   (c) 0
   (d) None of the above
       ANSWER:
       (a)

4. What is the vertical intercept of $h(t) = -f(t)$, If $f(t) = 2e^{0.07t}$? Answer without using a calculator or computer.

   (a) 2
   (b) -2
   (c) 0
   (d) None of the above
       ANSWER:
       (b)

For Problems 5–10, identify the statement as (a) true or (b) false.

5. The graph of the absolute value function is symmetric about the y-axis.
   ANSWER:
   (a) True

6. The two functions $f(x)$ and $f(-x)$ always have the same zeros.
   ANSWER:
   (b) False

7. The two functions $f(x)$ and $-f(x)$ always have the same zeros.
   ANSWER:
   (a) True

8. The functions $y = f(x)$ and $y = -f(x)$ have the same y-intercept.
   ANSWER:
   (b) False

9. The functions $y = f(x)$ and $y = f(-x)$ have the same y-intercept.
   ANSWER:
   (a) True

10. There is a function, f, such that $f(x) = f(-x) = -f(x) = -f(-x)$ for all x.
    ANSWER:
    (a) True, $f(x) = 0$
    COMMENT:
    Ask if there are others. Also, notice that $f(x) = f(-x) = -f(x)$ implies $f(x) = -f(-x)$.

Identify the functions in Problems 11–19 as:
(a) Odd
(b) Even
(c) Neither

11. $y = |x|$
    ANSWER:
    (b)

12. $y = |x| + 3$
    ANSWER:
    (b)

13. $y = |x + 3|$
    ANSWER:
    (c)

14. $y = 31$
    ANSWER:
    (b)

15. $y = e^{0.05x}$
    ANSWER:
    (c)

16. $y = -1/x$
    ANSWER:
    (a)

17. $y = (x + 1)(x - 3)$
    ANSWER:
    (c)

18. $y = x^{17}$
    ANSWER:
    (a)

19. $y = x^{17} + 3$
    ANSWER:
    (c)

For each table in Problems 20–23 state whether the given function could be

(a) Even or odd

(b) Even, but not odd

(c) Odd, but not even

(d) Neither odd nor even

20.

Table 5.1

x	-15	-10	-5	5	10	15
$f(x)$	-1	-3	-5	-5	-3	-1

ANSWER:
(b)

21.

Table 5.2

x	-5	-3	0	1	2	3
$f(x)$	7	5	0	-3	-4	-5

ANSWER:
(c)

22.

Table 5.3

x	-5	-3	-1	2	4	6
$f(x)$	5	3	1	2	4	6

ANSWER:
(a), there are no values to check.
COMMENT:
This looks like data from the absolute value function but are we sure?

23.

Table 5.4

x	-5	-3	0	3	5	6
$f(x)$	-5	-3	1	3	5	6

ANSWER:
(d)
COMMENT:
Discuss: If an odd function, f, is defined at $x = 0$, then $f(0) = 0$.

ConcepTests and Answers and Comments for Section 5.3

1. What is the value of k in the function $f(x) = k \cdot |x|$ as shown in Figure 5.3?

    (a) $3/2$
    (b) $2/3$
    (c) $-3/2$
    (d) $-2/3$

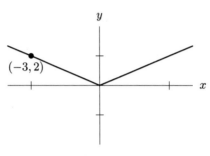

Figure 5.3

ANSWER:
(b)

2. Match the function graphs, of functions A to D, in Figure 5.3 to the following definitions:

    (I) $f(x) = \frac{1}{2}e^x$
    (II) $u(x) = -\frac{1}{2}e^x$
    (III) $v(x) = 2e^x$
    (IV) $w(x) = -2e^x$

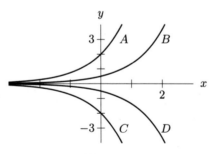

Figure 5.4

ANSWER:
    (I) B
    (II) D
    (III) A
    (IV) C

3. Match the following functions to the graphs in Figure 5.5

(I) $f(x) = (1/2)|x|$

(II) $u(x) = -2|x|$

(III) $v(x) = 2|x|$

(IV) $w(x) = -(1/2)|x|$

(a)
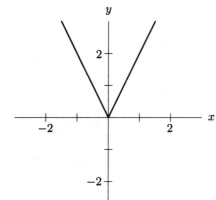

(b)
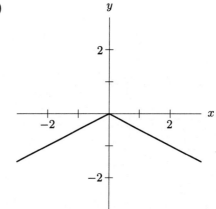

(c)
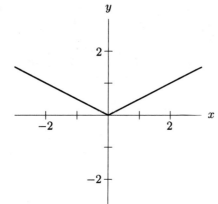

(d)
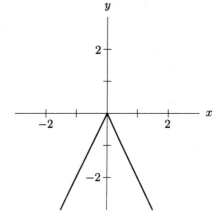

Figure 5.5

ANSWER:

(I) c

(II) d

(III) a

(IV) b

4. Match the following functions to the graphs in Figure 5.6

   (I) $v(x) = 2 - 3\ln(x - 1)$
   (II) $w(x) = 2 + 3\ln(x + 1)$
   (III) $t(x) = 2 + \ln(x + 1)$
   (IV) $u(x) = 2 - \ln(x - 1)$

(a)
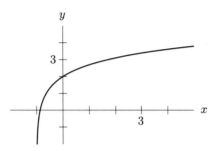

(b)
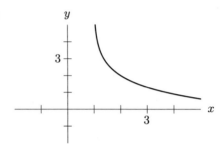

(c)
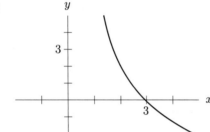

(d)
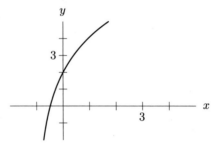

Figure 5.6

ANSWER:

(I) c
(II) d
(III) a
(IV) b

5. Use Table 5.5 to identify the value of k such that $f(x) = k \cdot g(x)$.

   (a) No possible k (b) 2   (c) -2 (d) $-\frac{1}{2}$ (e) $\frac{1}{2}$

Table 5.5

x	1	2	3	4
$f(x)$	-4	0	4	8
$g(x)$	2	0	-2	-4

ANSWER:
(c)

6. Use Table 5.6 to identify a value of k such that $f(x) = k \cdot g(x)$.

   (a) No possible k (b) 2   (c) -2 (d) $-\frac{1}{2}$ (e) $\frac{1}{2}$

Table 5.6

x	1	2	3	4
$f(x)$	-4	0	4	8
$g(x)$	-8	0	8	4^2

ANSWER:
(e)

7. Use Table 5.7 to identify a value of k such that $f(x) = k \cdot g(x)$.

   (a) No possible k

   (b) 3

   (c) -3

   (d) 1

Table 5.7

x	1	2	3	4
$f(x)$	$1 + \ln(1)$	$1 + \ln(2)$	$1 + \ln(3)$	$1 + \ln(4)$
$g(x)$	$1 + 3\ln(1)$	$1 + 3\ln(2)$	$1 + 3\ln(3)$	$1 + 3\ln(4)$

ANSWER:

(a)

8. What is the relationship between the two dotted lines shown in Figure 5.7?

   (a) one is twice as long as the other

   (b) one has twice the slope of the other

   (c) both (a) and (b)

   (d) neither (a) or (b)

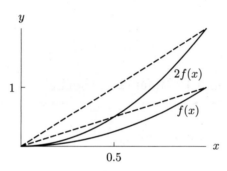

Figure 5.7

ANSWER:

(b)

COMMENT:

Note that the steeper line has twice the rise for the same run.

9. Determine the value of k in Figure 5.8.

   (a) 0.5

   (b) 0.75

   (c) 1.5

   (d) 3.0

   (e) None of the above

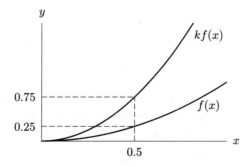

Figure 5.8

ANSWER:

(d)

10. Determine the value of k in Figure 5.9.

    (a) c/a

    (b) c/b

    (c) b/c

    (d) a/c

    (e) None of the above

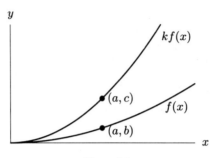

Figure 5.9

ANSWER:

(b)

ConcepTests and Answers and Comments for Section 5.4

For Problems 1–5 assume $k \neq 0$. Then identify the statement as

(a) Always true

(b) Sometimes true

(c) Never true

1. The functions $f(x)$ and $f(k \cdot x)$ have the same zeros.
   ANSWER:
   (b) Sometimes

2. The functions $f(x)$ and $f(k \cdot x)$ have the same vertical intercepts.
   ANSWER:
   (a) Always

3. $k \cdot f(x) = f(k \cdot x)$, for all x.
   ANSWER:
   (b) Sometimes
   COMMENT:
   Give an example for which this statement is true and one for which the statement is false.

4. $f(x) + k = f(k \cdot x)$, for all x.
   ANSWER:
   (c) Never
   COMMENT:
   Ask for justification, such as. if $x = 0$ then $k = 0$, which is a contradiction.

5. If the graph of the function $f(x)$ is concave up then the graph of $f(k \cdot x)$ is concave up.
   ANSWER:
   (a) Always
   COMMENT:
   Ask students about the effect on the graph when you multiply the input by k.

6. If $f(x) = e^x + 1$ and $g(x) = f(-2x)$ then which of the following is true?

   (a) $g(x) = -2xe^x + 1$
   (b) $g(x) = -2e^x + 1$
   (c) $g(x) = e^x - 2x + 1$
   (d) $g(x) = e^{-2x} + 1$
   (e) None of the above

   ANSWER:

   (d)

   COMMENT:

   This is a good questions to review input and output of a function. It is also good preparation for function composition.

7. If $f(x) = x^2$ and $g(x) = f(-2x)$, then which of the following is true?

   (a) $g(x) = 4f(x)$
   (b) $g(x) = -4x^2$
   (c) $g(x) = -2f(x)$
   (d) $g(x) = 4x$
   (e) None of the above

   ANSWER:

   (a)

   COMMENT:

   This is a good questions to review input and output of a function. It is also good preparation for function composition.

8. If $f(x) = x^2 + 3x - 4$ and $g(x) = f(-0.5x)$, then which of the following is true?

   (a) $g(x) = (1/4)x^2 + (3/2)x - 4$
   (b) $g(x) = -(1/4)x^2 - 3x - 4$
   (c) $g(x) = -(1/4)x^2 - (3/2)x - 4$
   (d) $g(x) = (1/4)x^2 - (3/2)x - 4$
   (e) None of the above

   ANSWER:

   (d)

   COMMENT:

   This is a good questions to review input and output of a function. It is also good preparation for function composition.

9. Match the functions below to the graphs, a–d, in Figure 5.10.

(I) $t(x) = 2\ln(x)$
(II) $r(x) = \ln(x + 2)$
(III) $u(x) = \ln(x) + 2$
(IV) $s(x) = \ln(2x)$

(a)
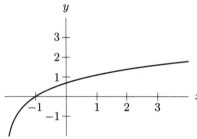

(b)
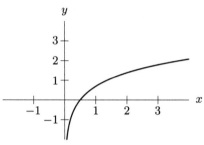

(c)
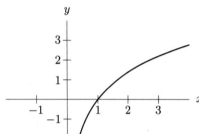

(d)
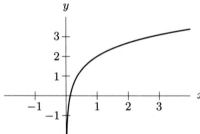

Figure 5.10

ANSWER:

(I) c
(II) a
(III) d
(IV) b

ConcepTests and Answers and Comments for Section 5.5

For Problems 1–6, identify the statement as (a) true or (b) false.

1. The standard form for a quadratic function makes it easy to identify the vertical intercept.
   ANSWER:
   (a) True
   COMMENT:
   What do the other forms make it easy to identify? Note that vertex form makes it easy to find zeros, in particular when the quadratic does not factor.

2. The graph of a quadratic function has two vertical asymptotes.
   ANSWER:
   (b) False

3. A quadratic function has a maximum and a minimum.
   ANSWER:
   (b) False
   COMMENT:
   What if the word "and" is changed to "or"?

4. The vertex form of a quadratic function can always be algebraically converted to standard form.
   ANSWER:
   (a) True

5. The standard form of a quadratic function can always be algebraically converted to vertex form.

ANSWER:

(a) True

COMMENT:

Which conversion direction is generally more difficult?

6. If $a < 0$, then the graph of the parabola, $y = -ax^2$ opens downward.

ANSWER:

(b) False

7. When you complete the square on the function $f(x) = x^2 + 6x + 11$, the squared term is?

(a) $(x - 3)^2$

(b) $(x + 3)^2$

(c) $(x - 6)^2$

(d) $(x + 6)^2$

(e) None of the above

ANSWER:

(b)

8. When you complete the square on the function $f(x) = x^2 + 6x + 11$, the amount to add and subtract is?

(a) $\sqrt{6}$ (b) 3   (c) 6   (d) 9   (e) None of the above

ANSWER:

(d)

For Problems 9–17, classify the equation as

(a) Quadratic in standard form

(b) Quadratic in vertex form

(c) Quadratic in factored form

(d) None of the above

9. $y = -x^2 + 3x + 2$

ANSWER:

(a)

10. $y = (x + 2) + 3$

ANSWER:

(d)

11. $y = (x + 2)(x - 1) + 3$

ANSWER:

(d)

12. $y = (x - \sqrt{2})(x + \sqrt{2})$

ANSWER:

(c)

13. $y = -(x + 3)^2 + 1$

ANSWER:

(b)

14. $y = 3x^2 + 4$

ANSWER:

Both (a) and (b)

15. $y = x^2 + \sqrt{3}x$

ANSWER:

(a)

16. $y = \sqrt{3x^2} + x$

ANSWER:

(d)

17. $y = 2^x + 3x + 4$

ANSWER:

(d)

For Problems 18–22, identify the vertex as either

(a) $(2, 4)$

(b) $(-2, 4)$

(c) $(2, -4)$

(d) $(-2, -4)$

(e) None of the above

18. $y = -3(x - 2)^2 + 4$
    ANSWER:
    (a)

19. $y = 4(x + 2)^2 - 4$
    ANSWER:
    (d)

20. $y = (x + 2)^2 + 4$
    ANSWER:
    (b)

21. $y = 5(x + 2)^2 + 4$
    ANSWER:
    (b)

22. $y = (x - 2)^2 + 4$
    ANSWER:
    (a)

23. Which of the following are possible equations for the function whose graph is given in Figure 5.11?

    (a) $y = (x - b)^2 + d$
    (b) $y = (x - b)^2 - d$
    (c) $y = (x - a)(x - c)$
    (d) $y = ax^2 + bx + c$
    (e) None of the above

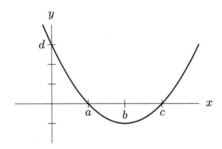

Figure 5.11

ANSWER:
(c)

24. Which of the following are possible equations for the function whose graph is given in Figure 5.12?

    (a) $y = -x(x-3) + 2.7$
    (b) $y = -1.2(x-1.5)^2 + 2.7$
    (c) $y = 1.2(x-1.5)^2 + 2.7$
    (d) $y = -x^2 + x$
    (e) None of the above

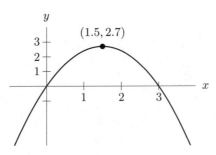

Figure 5.12

    ANSWER:

    (b)

25. Which of the following are possible equations for the function whose graph is given in Figure 5.13?

    (a) $y = x^2 - 3x$
    (b) $y = x^2 + 3x$
    (c) $y = (x-1.5)^2$
    (d) $y = (x+1.5)^2$
    (e) None of the above

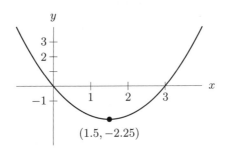

Figure 5.13

    ANSWER:

    (a)

26. If $y = x^2 + bx + c$ then to complete the square you add and subtract which one of the following values?

    (a) $b/2$
    (b) b/c
    (c) $(b/2)^2$
    (d) $\sqrt{b/2}$
    (e) None of the above

    ANSWER:

    (c)

27. Match the following quadratic functions to their vertex point.

(a) $f(x) = x^2 - 1$

(b) $u(x) = x^2 + 1$

(c) $v(x) = (x + 1)^2$

(d) $w(x) = (x - 1)^2$

(I) $(0, 1)$ (II) $(1, 0)$ (III) $(0, -1)$ (IV) $(-1, 0)$

ANSWER:

(a) III

(b) I

(c) IV

(d) II

Chapter Six

ConcepTests and Answers and Comments for Section 6.1 ━━━━━━━━

For Problems 1–4, identify the statement as (a) true or (b) false.

1. If f is a periodic function, then $f(t) = f(t) + c$, for all t.
   ANSWER:
   (b) False

2. A periodic function may have infinitely many zeros.
   ANSWER:
   (a) True

3. A periodic function may have only one zero.
   ANSWER:
   (b) False

4. As a bicycle rider starts a ride, the height of a trademark logo on the tire is a periodic function of time.
   ANSWER:
   (b) False, the time between successive zeros of the function decreases as speed increases.

For Problems 5–10, identify the statement as (a) true or (b) false. The ferris wheel height function gives the height above ground as a function of time. Unless otherwise noted, the wheel turns clockwise at a constant turning speed.

5. An increase in the diameter of a ferris wheel increases the period of the height function.
   ANSWER:
   (b) False

6. An increase in the diameter of a ferris wheel increases the amplitude of the height function.
   ANSWER:
   (a) True
   COMMENT:
   What is the relationship between the two?

7. If a ferris wheel is put on a raised platform then the amplitude of the height function increases.
   ANSWER:
   (b) False
   COMMENT:
   Ask about the midline.

8. The average rate of change of the height function over a one second interval is the same at the 1 o'clock position and the 3 o'clock position.
   ANSWER:
   (b) False
   COMMENT:
   Ask students which is greater and to explain why.

9. The average rate of change of the height function over a one second interval is positive at the 9 o'clock position and negative at the 3 o'clock position.
   ANSWER:
   (a) True
   COMMENT:
   Ask students if they are the same magnitude.

10. If a ferris wheel does not have a constant speed but stops one minute every ten minutes, the height function can not be periodic.
    ANSWER:
    (b) False
    COMMENT:
    How does the graph change?

For Problems 11–13, give the value of the designated property of the periodic function shown in Figure 6.1.

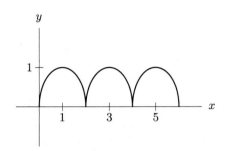

Figure 6.1

11. Period.

    (a) 1   (b) 0   (c) 1.25   (d) 5.5   (e) None of the above

    ANSWER:
    (e)

12. Amplitude.

    (a) 0.5   (b) 0   (c) 1.25   (d) 1.5   (e) None of the above

    ANSWER:
    (a)

13. Height of midline.

    (a) 0.5   (b) 0   (c) 1.25   (d) 1.5   (e) None of the above

    ANSWER:
    (a)

Rounding numbers can be thought of as adding an amount to round the given number to the closest integer. For example, rounding 6.3 to an integer requires adding -0.3. The amount added is a function $d(x)$ of the given number; for example, $d(6.3) = -0.3$, and $d(6.8) = 0.2$. A further sample of function values is given in Table 6.1. For Problems 14–16, find the designated value of the attribute, i.e. period, amplitude or midline, for the periodic function $d(x)$.

Table 6.1

x	1	1.2	1.9	2.49	2.5	2.51
$d(x)$	0	−0.2	0.1	−0.49	0.5	0.49

14. Period.

    (a) −0.5   (b) 0   (c) 0.5   (d) 1   (e) None of the above

    ANSWER:
    (d)

15. Amplitude.

    (a) −0.5   (b) 0   (c) 0.5   (d) 1   (e) None of the above

    ANSWER:
    (c)

16. Height of midline.

    (a) −0.5   (b) 0   (c) 0.5   (d) 1   (e) None of the above

    ANSWER:
    (b)

For Problems 17–19, give the value of the designated attribute, i.e. period, amplitude, or midline, of the periodic function shown in Figure 6.2.

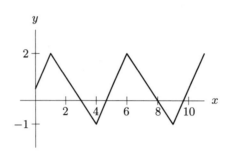

Figure 6.2

17. Period.

(a) 0.5   (b) 1   (c) 1.5   (d) 3   (e) 5

ANSWER:
(e)

18. Amplitude.

(a) 0.5   (b) 1   (c) 1.5   (d) 3   (e) 5

ANSWER:
(c)

19. Height of midline.

(a) 0.5   (b) 1   (c) 1.5   (d) 3   (e) 5

ANSWER:
(a)

20. Three students drew graphs to describe the ferris wheel height function over time. They made three different assumptions about the value of the height function during a stop. Student A assumed that the value of the height function would be zero, student B assumed the value of the height function would be a constant, and student C assumed the value of the height function would be undefined. Match the three assumption to graphs (a)–(c).

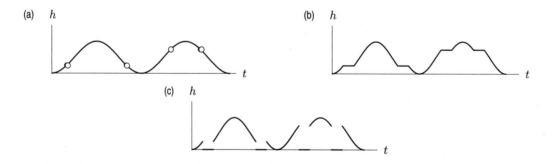

ANSWER:

(a) C
(b) B
(c) A

COMMENT:
Ask students why student B made the correct assumption.

ConcepTests and Answers and Comments for Section 6.2

For Problems 1–4, identify the statement as (a) true or (b) false.

1. The unit circle is one unit across.
    ANSWER:
    (b) False
    COMMENT:
    The radius is one unit.

2. Angles are measured in a clockwise direction on the unit circle.
    ANSWER:
    (b) False

3. When measured from the same point, the angles $210°$ and $-150°$ end at the same point on a unit circle.
    ANSWER:
    (a) True
    COMMENT:
    However, the angles are not equal.

4. The angles $210°$ and $-150°$ are equal.
    ANSWER:
    (b) False

5. How many of the following points are on the unit circle? $(0, -1)$, $(1, 1)$, $(1, 0)$, $(\frac{1}{2}, \frac{1}{2})$, $(0, 0)$, $(0, 1)$

    (a) 2
    (b) 3
    (c) 4
    (d) 5
    (e) None of the above

        ANSWER:
        (b)

6. For how many angles θ, between $0°$ and $360°$, does $\cos\theta = \sin\theta$?

    (a) 2
    (b) 3
    (c) 4
    (d) 5
    (e) None of the above

        ANSWER:
        (a)

7. When the hands on a clock show 1:30, the angle between the hands is:

    (a) $90°$
    (b) $125°$
    (c) $130°$
    (d) $135°$
    (e) $150°$

        ANSWER:
        (d)
        COMMENT:
        Note that the hand hours is half way between 1 and 2.

8. If the point $(1, 2)$ is on a circle centered at the origin, the circle has radius:

    (a) 1
    (b) 3
    (c) 5
    (d) $\sqrt{3}$
    (e) $\sqrt{5}$

        ANSWER:
        (e)

9. In what quadrant of the unit circle are both the sine and cosine values negative?

    (a) None
    (b) I
    (c) II
    (d) III
    (e) IV

    ANSWER:

    (d)

10. When measured from the horizontal axis on a unit circle, how many of the following angles terminate on a vertical or horizontal axis? $90°, 630°, 100°, -270°, 0°, 360°$

    (a) 2
    (b) 3
    (c) 4
    (d) 5
    (e) None of the above

    ANSWER:

    (d)

    COMMENT:

    Which angles determine the same point on the unit circle?

11. $P = (1/2, 1/2)$ is a point on a circle centered at the origin, O. If θ is the angle the segment $\overline{OP}$ makes with the positive x-axis, then

    (a) The circle is a unit circle
    (b) The angle θ has measure $45°$
    (c) Both (a) and (b)
    (d) Neither (a) or (b)

    ANSWER:

    (b)

ConcepTests and Answers and Comments for Section 6.3

For Problems 1–4, identify the statement as (a) true or (b) false.

1. The unit circle is useful for degree measures, but not for radian measure of angles.
    ANSWER:
    (b) False

2. Calculators can be used to convert degrees to radians.
    ANSWER:
    (a) True

3. $360°$ is a hot oven temperature.
    ANSWER:
    (b) False
    COMMENT:
    This question is a reminder about proper notation. It distinguishes $360°$ from $360°$F.

4. Degree measure is always an integer value.
    ANSWER:
    (b) False

5. Which is closest to $360°$?

    (a) 3 radians
    (b) 4 radians
    (c) 6 radians
    (d) 7 radians
    (e) None of the above

    ANSWER:
    (c)
    COMMENT:
    Using a string that has length equal to the radius of the circle works well to demonstrate this. Just trace the circumference of the circle with the string.

6. Which is closest to 2 radians?

    (a) $(\pi/90)°$
    (b) $60°$
    (c) $120°$
    (d) $180°$
    (e) $270°$

    ANSWER:
    (c)

7. Which of the following are defined?

    (a) $\pi°$
    (b) $\frac{1}{2}°$
    (c) $-0.5°$
    (d) $-270°$
    (e) All of the above

    ANSWER:
    (e)

8. The minute hand on a clock at the Musée d'Orsay in Paris is 3 meters in length. How far does the tip of the hand travel in one hour?

    (a) 3 m
    (b) 6 m
    (c) 3π m
    (d) 6π m
    (e) None of the above

    ANSWER:
    (d)

9. If the unit of measure is the diameter of a circle, then if we measure counterclockwise around the circumference of the circle, the angle formed has a measure of how many radians?

    (a) 2
    (b) 1
    (c) π
    (d) varies by circle radius
    (e) varies by starting point

    ANSWER:
    (a)

10. Order the size of the following angles from smallest to largest.

    (a) $30°$
    (b) 1
    (c) $\pi/2$
    (d) $\pi/4$
    (e) $60°$

    ANSWER:
    (a), (d), (b), (e), (c)

ConcepTests and Answers and Comments for Section 6.4 ━━━━━━

1. Match the graphs in Figure 6.3 to the following list of functions:

   (a) $y = \sin x$
   (b) $y = \cos x$
   (c) $y = 1 + \sin x$
   (d) $y = 1 + \cos x$
   (e) $y = \frac{3}{2} \sin x$
   (f) $y = \frac{3}{2} \cos x$

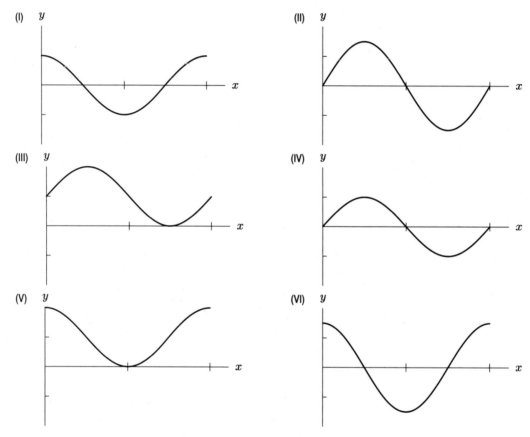

Figure 6.3: All graphs have equal scale

ANSWER:

   (I)   b
   (II) e
   (III) c
   (IV) a
   (V)   d
   (VI) f

2. In Figure 6.4, which of the following describes the relationship of the unit circle and the graph of the sine function.

   (a) They have the same axes
   (b) The vertical axis is the same but the horizontal axis is different
   (c) The horizontal axis is the same but the vertical axis is different
   (d) Neither axis is the same
   (e) There is no relation between the two graphs

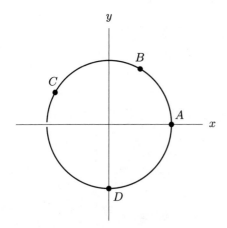 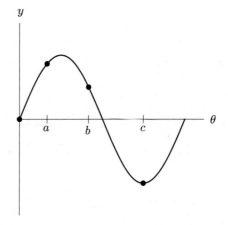

Figure 6.4: The unit circle and the sine function graph

ANSWER:
(b)

3. True or false: In Figure 6.4, there is a relationship between the four points named on the unit circle and the four point shown on the sine function graph.

   (a) True
   (b) False

   ANSWER:
   (a)
   COMMENT:
   Ask the students to articulate the relationship.

4. In Figure 6.4, the distance on the circle from A to C is equal to what value on the sine function graph?
   (a) a   (b) b   (c) c   (d) 0   (e) A value not shown
   ANSWER:
   (b)

For Problems 5–8, identify the midline as

(a) $y = -2$ (b) $y = -1$ (c) $y = 1$ (d) $y = 2$ (e) $y = 3$

5. $y = 2 + 3\sin t$
   ANSWER:
   (d)

6. $y = 3 - \sin t$
   ANSWER:
   (e)

7. $y = 3 - 2\sin t$
   ANSWER:
   (e)

8. $y = -2 + \sin t$
   ANSWER:
   (a)

For Problems 9–12, identify the amplitude as

(a) -2

(b) -1

(c) 1

(d) 2

(e) 3

9. $y = 2 + \sin t$
   ANSWER:
   (c)

10. $y = 3 - \sin t$
    ANSWER:
    (c)
    COMMENT:
    Note that amplitude is always positive.

11. $y = 3 - 2 \sin t$
    ANSWER:
    (d)

12. $y = 2 + 3 \sin t$
    ANSWER:
    (e)

ConcepTests and Answers and Comments for Section 6.5

1. The function $f(t) = 3 \sin 2t + C$ has no zeros if C is greater than

   (a) 0
   (b) 1
   (c) 2
   (d) 3
   (e) None of the above

   ANSWER:
   (d)
   COMMENT:
   What is the effect of adding C?

2. If $A_1 \sin(B_1 t) = A_2 \cos(B_2 t + C)$ for all values of t and A_1, A_2, B_1 and B_2 are all non-zero, then

   (a) $A_1 = A_2$
   (b) $B_1 = B_2$
   (c) $C = \frac{\pi}{2} \cdot n$, for some integer n
   (d) All of the above
   (e) None of the above

   ANSWER:
   (d)
   COMMENT:
   Actually, $C = -\frac{\pi}{2} + 2\pi n$ for some integer n.

3. The domain for the function $f(t) = A \sin(Bt + C) + D$ is altered by changing which of the following?

   (a) A
   (b) B
   (c) C
   (d) D
   (e) None of the above

   ANSWER:
   (e)

4. The range for the function $f(t) = A\sin(Bt + C) + D$ is altered by changing which of the following?

   (a) A
   (b) B and C
   (c) A and D
   (d) D
   (e) None of the above

   ANSWER:

   (c)

   COMMENT:

   If B and C are zero, then (d) is the only correct answer.

5. Comparing $f(t) = 2\sin(3t + \frac{\pi}{4})$ to $g(t) = 2\sin(3t)$, the phase shift is

   (a) 2
   (b) 3
   (c) $\frac{\pi}{12}$
   (d) $\frac{\pi}{4}$
   (e) None of the above

   ANSWER:

   (d)

   COMMENT:

   What is the horizontal shift?

6. The amplitude of the function $f(t) = \cos(5t + C) + \cos(5t)$ is the greatest for which value of C?

   (a) 0
   (b) $\frac{\pi}{5}$
   (c) $\frac{\pi}{4}$
   (d) $\frac{\pi}{2}$
   (e) π

   ANSWER:

   (a)

   COMMENT:

   What happens when C is zero?

7. The amplitude of the function $f(t) = \cos(5t + C) + \cos(5t)$ is the least for which value of C?

   (a) 0
   (b) $\frac{\pi}{10}$
   (c) $\frac{\pi}{5}$
   (d) $\frac{\pi}{2}$
   (e) π

   ANSWER:

   (e)

   COMMENT:

   What happens when $C = \pi$?

8. The period of the function $f(t) = \cos(\frac{t}{2} + \frac{\pi}{2})$ is

   (a) $\frac{\pi}{4}$
   (b) $\frac{\pi}{2}$
   (c) π
   (d) 2π
   (e) 4π

   ANSWER:

   (e)

9. The phase shift of the function $f(t) = \cos(\frac{t}{2} + \frac{\pi}{2})$ is

   (a) $\frac{\pi}{4}$

   (b) $\frac{\pi}{2}$

   (c) π

   (d) 2π

   (e) 4π

     ANSWER:

     (b)

     COMMENT:

     What is the horizontal shift?

10. In Figure 6.5 the solid graph is for $f(t) = \cos(2t)$. What is the equation for the dotted graph?

   (a) $y = \cos(2t + \frac{\pi}{4})$

   (b) $y = \cos(2t - \frac{\pi}{4})$

   (c) $y = \cos(2t + \frac{\pi}{2})$

   (d) $y = \cos(2t - \frac{\pi}{2})$

   (e) None of the above

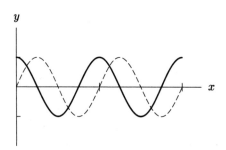

Figure 6.5

     ANSWER:

     (d)

     COMMENT:

     The phase shift from $\cos 2t$ is $\frac{1}{4}$ period, which is $(\frac{1}{4}\text{period})(2\pi\frac{\text{radians}}{\text{period}}) = \frac{\pi}{2}$ radians. A translation to the right gives $-\frac{\pi}{2}$ radians.

ConcepTests and Answers and Comments for Section 6.6 ──────────

1. If the graph of the function passes through the origin then the function could be

   (a) $\sin(Bt + C)$

   (b) $\cos(Bt + C)$

   (c) $\tan(Bt + C)$

   (d) All of the above

   (e) None of the above

     ANSWER:

     (d)

     COMMENT:

     What does C have to be in each case?

2. The domain of the tangent function is

   (a) All reals
   (b) All reals except odd integer multiples of $\frac{\pi}{2}$
   (c) $-\frac{\pi}{2} \leq y \leq \frac{\pi}{2}$
   (d) $-\frac{\pi}{2} < y < \frac{\pi}{2}$
   (e) None of the above

   ANSWER:
   (b)
   COMMENT:
   Ask students why this is the case.

3. The range of the tangent function is

   (a) All reals
   (b) $-1 \leq y \leq 1$
   (c) $-\frac{\pi}{2} \leq y \leq \frac{\pi}{2}$
   (d) $-\frac{\pi}{2} < y < \frac{\pi}{2}$
   (e) None of the above

   ANSWER:
   (a)

4. The graph of the function $y = \tan x$ and the graph of $y = x$ intersect in how many points?

   (a) 0
   (b) 1
   (c) 2
   (d) 3
   (e) infinitely many

   ANSWER:
   (e)

5. If $\tan \theta = 5$ then the value of $\cot \theta =$ is

   (a) -5
   (b) -1/5
   (c) 1/5
   (d) 5
   (e) None of the above

   ANSWER:
   (c)

6. The function $f(t) = 3 \tan 2t + C$ has no zeros if C is greater than

   (a) 0
   (b) 1
   (c) 2
   (d) 3
   (e) None of the above

   ANSWER:
   (e)

7. The range of the secant function is

   (a) All reals
   (b) All $y, |y| \geq 1$
   (c) $-\frac{\pi}{2} \leq y \leq \frac{\pi}{2}$
   (d) $-\frac{\pi}{2} < y < \frac{\pi}{2}$
   (e) None of the above

   ANSWER:
   (b)
   COMMENT:
   Ask students why this is the case.

8. The domain of the secant function is

   (a) All reals
   (b) All y, $|y| \geq 1$
   (c) All reals, except odd integer multiples of $\frac{\pi}{2}$
   (d) All reals, except even integer multiples of $\frac{\pi}{2}$
   (e) None of the above

   > ANSWER:
   > (c)
   > COMMENT:
   > Why is this the same as for the tangent function?

9. The range of the cosecant function is

   (a) All reals
   (b) All y, $|y| \geq 1$
   (c) $-\frac{\pi}{2} \leq y \leq \frac{\pi}{2}$
   (d) $-\frac{\pi}{2} < y < \frac{\pi}{2}$
   (e) None of the above

   > ANSWER:
   > (b)
   > COMMENT:
   > Ask students why this is the case.

10. The domain of the cosecant function is

    (a) All reals
    (b) All y, $|y| \geq 1$
    (c) All reals, except odd integer multiples of $\frac{\pi}{2}$
    (d) All reals, except integer multiples of π
    (e) None of the above

    > ANSWER:
    > (d)

11. The value of $\sec(30°)$ is

    (a) $\frac{1}{2}$
    (b) $\frac{\sqrt{3}}{2}$
    (c) $\frac{2}{\sqrt{3}}$
    (d) 2
    (e) None of the above

    > ANSWER:
    > (c)

12. The graphs of $\csc t$ and $\sec t$ have what relationship?

    (a) None
    (b) Same domain and range
    (c) One is a horizontal shift of the other
    (d) One is a vertical shift of the other
    (e) One is the reciprocal of the other

    > ANSWER:
    > (c)
    > COMMENT:
    > Ask students why this is the case.

ConcepTests and Answers and Comments for Section 6.7

1. The inverse sine and inverse cosine functions have

    (a) The same domain
    (b) The same range
    (c) Both (a) and (b)
    (d) Neither (a) and (b)

    ANSWER:
    (a)

2. The arccosine function has the range

    (a) $0 \leq t \leq \pi$
    (b) $-\pi/2 \leq t \leq \pi/2$
    (c) All reals
    (d) $t > 0$
    (e) None of the above

    ANSWER:
    (a)

3. The arcsin t and arctan t functions have the same range of

    (a) $0 \leq t \leq \pi$
    (b) $-\pi/2 \leq t \leq \pi/2$
    (c) All reals
    (d) $t > 0$
    (e) None of the above

    ANSWER:
    (e)
    COMMENT:
    Review $-\pi/2 \leq t \leq \pi/2$ versus $-\pi/2 < t < \pi/2$, if needed to see the difference between the two domains.

4. For how many values of t is arcsin t = arctan t?

    (a) none
    (b) 1
    (c) 2
    (d) 3
    (e) Infinitely many

    ANSWER:
    (b)

5. Which of the following are equal to $1/\sin t$.

    (a) $\sin^{-1} t$
    (b) $\sin t^{-1}$
    (c) $(\sin t)^{-1}$
    (d) All of the above
    (e) None of the above

    ANSWER:
    (c)

6. For how many t does $\sin^{-1} t = \cos^{-1} t$.

    (a) 0
    (b) 1
    (c) 2
    (d) 3
    (e) infinitely many

    ANSWER:
    (b)

7. Which of the following sets includes only reference angles.

   (a) $\{0, \pi, 2\pi, 3\pi, 4\pi, 5\pi\}$
   (b) $\{1°, 2°, 3°, 4°, 5°\}$
   (c) $\{0, \pi/16, \pi/8, \pi/4, \pi/2, \pi\}$
   (d) $\{-\pi/4, \pi/4, -\pi/3, \pi/3, -\pi/6, \pi/6\}$
   (e) None of the above

   ANSWER:
   (b)

8. The range of $\sin^{-1} t$ is:

   (a) All reals
   (b) $-1 \leq t \leq 1$
   (c) $-\frac{\pi}{2} \leq t \leq \frac{\pi}{2}$
   (d) $0 \leq t \leq \pi$
   (e) None of the above

   ANSWER:
   (c)

9. The range of $\cos^{-1} t$ is:

   (a) All reals
   (b) $-1 \leq t \leq 1$
   (c) $-\frac{\pi}{2} \leq t \leq \frac{\pi}{2}$
   (d) $0 \leq t \leq \pi$
   (e) None of the above

   ANSWER:
   (d)

10. The domain of $\tan^{-1} t$ is:

    (a) All reals
    (b) $-1 \leq t \leq 1$
    (c) $-\frac{\pi}{2} \leq t \leq \frac{\pi}{2}$
    (d) $0 \leq t \leq \pi$
    (e) None of the above

    ANSWER:
    (a)

11. The range of $\tan^{-1} t$ is:

    (a) All reals
    (b) $-\frac{\pi}{2} < t < \frac{\pi}{2}$
    (c) $-\frac{\pi}{2} \leq t \leq \frac{\pi}{2}$
    (d) $0 < t < \pi$
    (e) None of the above

    ANSWER:
    (b)

Chapter Seven

Chapter Seven

ConcepTests and Answers and Comments for Section 7.1 ━━━━

For Problems 1–11, use Figure 7.1 to identify the statement as (a) true, (b) false or (c) unknown.

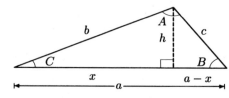

Figure 7.1

1. $\frac{1}{2}x = \frac{x}{2}$
   ANSWER:
   (a) True
   COMMENT:
   This question has nothing to do with the figure, but it is still good to include as a warm-up.

2. $a > b$
   ANSWER:
   (a) True

3. A right angle is shown in the figure.
   ANSWER:
   (a) True

4. An acute angle is shown in the figure.
   ANSWER:
   (a) True

5. An obtuse angle is shown in the figure.
   ANSWER:
   (a) True

6. $a^2 + b^2 = c^2$
   ANSWER:
   (b) False
   COMMENT:
   You may want to ask students if they can apply the Pythagorean Theorem to any side lengths.

7. $a + b + c = 180$
   ANSWER:
   (c) Unknown

8. $A + B + C = 180°$
   ANSWER:
   (a) True

9. $C < 90°$
   ANSWER:
   (a) True

10. $\cos C = \frac{x}{b}$
    ANSWER:
    (a) True

11. $\cos B = \frac{a-x}{h}$
    ANSWER:
    (b) False
    COMMENT:
    Note that $\cos B = \frac{a-x}{c}$.

For Problems 12–16, three elements of the triangle in Figure 7.2 are given. Decide whether (a) the Law of Cosines, (b) the Law of Sines, or (c) neither of these can be used to solve for the unknown sides and angles.

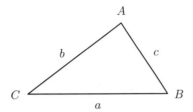

Figure 7.2

12. a, b, c

    ANSWER:

    (a) Law of Cosines

    COMMENT:

    Since all sides are known we can use the Law of Cosines to find any of the angles.

13. A, B, C

    ANSWER:

    (c) Neither

14. b, C, a

    ANSWER:

    (a) Law of Cosines

15. C, a, B

    ANSWER:

    (b) Law of Sines

    COMMENT:

    Even though angle A is not given, we can compute it in this case.

16. A, a, B

    ANSWER:

    (b) Law of Sines

17. In the ambiguous case for solving triangles,

    (a) All three sides have the same length
    (b) All three angles have the same measure
    (c) One side may have two different lengths
    (d) One angle may have two different measures
    (e) None of the above

    ANSWER:

    (d)

ConcepTests and Answers and Comments for Section 7.2 ━━━━━━━

For Problems 1–12, answer (a) true if the equation is an identity or (b) false if it is not an identity.

1. $\frac{1}{2}x = \frac{x}{2}$
   ANSWER:
   (a) True

2. $\tan y = \frac{\cos y}{\sin y}$
   ANSWER:
   (b) False
   COMMENT:
   Ask the students to correct the equation to make an identity.

3. $\sin^2 \frac{\theta}{2} + \cos^2 \frac{\theta}{2} = 1$
   ANSWER:
   (a) True

4. $\frac{1}{2}\sin^2 x + \frac{1}{2}\cos^2 x = 1$
   ANSWER:
   (b) False

5. $\sin^2(x+1) + \cos^2(x+1) = 1$
   ANSWER:
   (a) True

6. $(\sin\theta + \cos\theta)^2 = 1$
   ANSWER:
   (b) False

7. $\cos 2\theta = 2\sin\theta\cos\theta$
   ANSWER:
   (b) False

8. $\sin 2t = 2\sin t\cos t$
   ANSWER:
   (a) True

9. $\cos 2x = 1 - \sin^{-1} x$
   ANSWER:
   (b) False

10. $\sin(-t) = -\sin t$
    ANSWER:
    (a) True
    COMMENT:
    Recall that $y = \sin t$ is an odd function.

11. $\cos(-t) = -\cos t$
    ANSWER:
    (b) False
    COMMENT:
    Recall that $y = \cos t$ is an even function, thus $\cos(-t) = \cos t$.

12. $\tan(-t) = -\tan t$
    ANSWER:
    (a) True

13. Without a calculator or computer, match the following functions to the graphs in Figure 7.3.

    (I) $y = \cos t$
    (II) $y = \cos 2t$
    (III) $y = \sin t$
    (IV) $y = \sin 2t$
    (V) $y = 2\sin t \cos t$
    (VI) $y = \cos(t - \frac{\pi}{2})$
    (VII) $y = \cos^2 t - \sin^2 t$
    (VIII) $y = \sin(t + \frac{\pi}{2})$

(a)
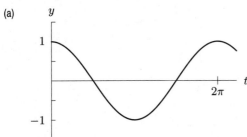

(b)
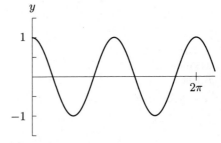

(c)
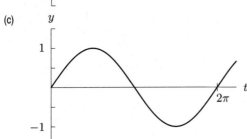

(d)
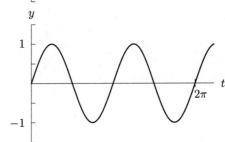

Figure 7.3

ANSWER:

    (I) (a)
    (II) (b)
    (III) (c)
    (IV) (d)
    (V) (d)
    (VI) (c)
    (VII) (b)
    (VIII) (a)

14. Without a graphing utility, match the following functions to the graphs in Figure 7.4.

(I) $y = 2\sin t \cos t$
(II) $y = \cos(-t)$
(III) $y = \cos^2 t + \sin^2 t$
(IV) $y = -\sin t / \cos t$

(a)
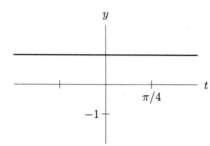

(b)
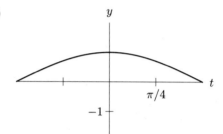

(c)
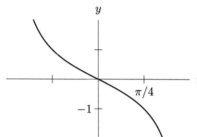

(d)
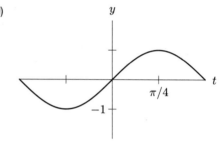

Figure 7.4

ANSWER:

(I)   (d)
(II) (b)
(III) (a)
(IV) (c)

ConcepTests and Answers and Comments for Section 7.3

For Problems 1–2, decide whether the statement is (a) true or (b) false?

1. An identity can contain only one variable.
   ANSWER:
   (b) False

2. True or false: $\cos u + \cos v = \cos(u + v)$, for all u and v.
   ANSWER:
   (b) False

3. How many of the following items have known exact values because they are related to special angles?
   $\cos 30°$, $\sin 45°$, $\sin 405°$, $\cos 315°$, $\tan -45°$

   (a) 0
   (b) 2
   (c) 3
   (d) 4
   (e) 5

   ANSWER:
   (e)

4. In the following list, how many items have known exact values, because they are related to special angles?
$\cos 2\pi, \sin \pi/2, \sin 2\pi/3, \cos 5\pi/7, \tan 1$

(a) 0   (b) 2   (c) 3   (d) 4   (e) 5

ANSWER:

(c)

5. How many of the following expressions can be written in terms of $\cos A, \sin A, \cos B, \sin B$?
$\sin(A + B), \sin(A * B), \sin(A - B), \sin(A/B), \tan(A + B)$

(a) 0

(b) 2

(c) 3

(d) 4

(e) 5

ANSWER:

(c)

6. Which of the following expressions can be written as a single sine function of the form $A\sin(Bt + \theta)$?

(I) $3\sin(2t) + 5\cos(2t)$

(II) $\cos(\pi t)$

(III) $2\sin(3t) + 3\cos(2t)$

(IV) $\sin(2t) + \tan(2t)$

(V) $\cos(\pi t) - 5\sin(\pi t)$

(a) II only

(b) I, II, III, V

(c) I, II, V

(d) All of them

ANSWER:

(c)

7. If given the values of $\cos 2$ and $\cos 3$, $\sin 2$ and $\sin 3$, which of the following can be found arithmetically?

(I) $\cos(1)$

(II) $\cos(5)$

(III) $\sin(-1)$

(IV) $\cos(\pi + 2)$

(V) $\tan(4)$

(a) I, II, V

(b) I, II, III

(c) I, II, III, V

(d) All of them

ANSWER:

(d)

8. If given the values of $\cos 2$ and $\cos 3$, $\sin 2$ and $\sin 3$, which of the following can be found arithmetically?

(I) $\cos 6$

(II) $\cos 1.5$

(III) $\sin 4$

(IV) $\tan 1$

(V) $\cot 4$

(a) I, III

(b) I, III, V

(c) I, II, III, V

(d) All of them

ANSWER:

(d)

COMMENT:

Assumes that students know that there are half-angle formulas. They are in the exercises. Alternatively, students can use the double angle formula $\cos 3 = \cos(2(1.5)) = 2\cos^2(1.5) - 1$ which can be solved for $\cos 1.5$.

ConcepTests and Answers and Comments for Section 7.4 ━━━━━━━━

1. Match the functions (I) - (III) for use in a model with period and time units listed as (a) - (c)

   (I) $Q = a_1 \sin(\frac{2\pi}{12} t)$ (II) $Q = a_2 \sin(\frac{\pi}{12} t)$ (III) $Q = a_4 \sin(\frac{\pi}{30} t)$

   (a) 1 day, with time in hours
   (b) 1 hour, with time in minutes
   (c) 1 year, with time in months

   ANSWER:

   (I) c
   (II) a
   (III) b

2. The amplitude of the function $4\cos(t) + 5\cos(t + \pi)$ is

   (a) 9
   (b) 4.5
   (c) 1
   (d) $\sqrt{4^2 + 5^2}$
   (e) None of the above

   ANSWER:
   (c)
   COMMENT:
   Note that $4\cos(t) + 5\cos(t + \pi) = -\cos(t)$.

3. The period of the function $4\cos(t) + 5\cos(t + \pi)$ is

   (a) π
   (b) 2π
   (c) 1
   (d) 2
   (e) None of the above

   ANSWER:
   (b)

4. The period of the function $4\cos(2t) + 5\cos(4t)$ is

   (a) π
   (b) 2π
   (c) $\pi/2$
   (d) 2
   (e) None of the above

   ANSWER:
   (a)
   COMMENT:
   Ask students to generalize this result.

5. Which of the following functions exhibit damped oscillation?

   (a) $f(t) = \cos(e^t)$
   (b) $g(t) = t\cos(2\pi t)$
   (c) $h(t) = \frac{1}{1000}\sin(2\pi t)$
   (d) $m(t) = e^{-3t}\tan t$
   (e) None of the above

   ANSWER:
   (e)
   COMMENT:
   Ask students for an example of damped oscillation.

ConcepTests and Answers and Comments for Section 7.5

For Problems 1–6, is the statement (a) true or (b) false?

1. Polar coordinates of a given point are unique.
   ANSWER:
   (b) False

2. Cartesian coordinates of a given point are unique.
   ANSWER:
   (a) True

3. Conversion from Cartesian coordinates to polar coordinates requires trigonometry.
   ANSWER:
   (a) True

4. For points $P_1 = (r_1, \theta_1)$ and $P_2 = (r_2, \theta_2)$, if $r_1 \neq r_2$, then $P_1 \neq P_2$.
   ANSWER:
   (a) True
   COMMENT:
   Recall, all values of r are non-negative.

5. For points $P_1 = (r_1, \theta_1)$ and $P_2 = (r_2, \theta_2)$, if $\theta_1 \neq \theta_2$, then $P_1 \neq P_2$.
   ANSWER:
   (b) False

6. The equation of the unit circle centered at the origin has the polar equation $r = 1$.
   ANSWER:
   (a) True

7. The shaded region in Figure 7.5 is described by

   (a) $-2 \leq r \leq -1$ and $0 \leq \theta \leq \frac{\pi}{2}$
   (b) $-2 \leq r \leq -1$ and $\frac{\pi}{2} \leq \theta \leq \pi$
   (c) $1 \leq r \leq 2$ and $0 \leq \theta \leq \frac{\pi}{2}$
   (d) $1 \leq r \leq 2$ and $\frac{\pi}{2} \leq \theta \leq \pi$
   (e) None of the above

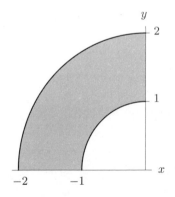

Figure 7.5

ANSWER:
(d)

8. The shaded region in Figure 7.5 is described in Cartesian coordinates by
   (a) $-2 \leq x \leq -1$ and $0 \leq y \leq \sqrt{4 - x^2}$, and, $-1 \leq x \leq 0$ and $\sqrt{1 - x^2} \leq y \leq \sqrt{4 - x^2}$
   (b) $-2 \leq x \leq -1$ and $\sqrt{1 - x^2} \leq y \leq \sqrt{4 - x^2}$, and, $-1 \leq x \leq 0$ and $\sqrt{1 - x^2} \leq y \leq \sqrt{4 - x^2}$
   (c) $-2 \leq x \leq 0$ and $\sqrt{1 - x^2} \leq y \leq \sqrt{4 - x^2}$
   (d) $-2 \leq x \leq 0$ and $0 \leq y \leq \sqrt{4 - x^2}$
   (e) None of the above

   ANSWER:
   (a)
   COMMENT:
   This question is meant as a comparison with the easier description of the region in polar coordinates.

ConcepTests and Answers and Comments for Section 7.6

For Problems 1–10, is the statement (a) true or (b) false?

1. The conjugate of a real number is the opposite of itself.
   ANSWER:
   (b) False

2. Zero is the only purely imaginary number that is its own conjugate.
   ANSWER:
   (a) True

3. Adding two complex numbers can result in a purely imaginary number.
   ANSWER:
   (a) True

4. Adding two complex numbers can result in a real number.
   ANSWER:
   (a) True
   COMMENT:
   Especially since any real number is also a complex number.

5. Division is not possible for complex numbers.
   ANSWER:
   (b) False
   COMMENT:
   Note the restriction to a non-zero divisor.

6. The number i is real and irrational.
   ANSWER:
   (b) False

7. For complex numbers, $(a + bi)^2 = a^2 + ib^2$.
   ANSWER:
   (b) False
   COMMENT:
   What about $(a + bi)^2 = a^2 + (ib)^2$? That is still wrong, of course.

8. $e^{i\pi} = -1$.
   ANSWER:
   (a) True

9. The product of a complex number and its conjugate is a real number.
   ANSWER:
   (a) True

10. The sum of a complex number and its conjugate is a real number.
    ANSWER:
    (a) True

11. Which complex number is represented by a point with polar coordinates $r = 2$ and $\theta = \pi/4$?

    (a) $2 + (\pi/4)i$
    (b) $2\cos(\pi/4) + 2\sin(\pi/4)$
    (c) $\cos 2 + i\sin(\pi/4)$
    (d) $\sqrt{2} + i\sqrt{2}$
    (e) None of the above

    ANSWER:
    (d)

12. The point with Cartesian coordinates $(3, 4)$ corresponds to which point in the complex plane?

    (a) $z = 3 + 4i$
    (b) $z = 4 + 3i$
    (c) $z = 9 + 16i$
    (d) $z = 3e^{4i}$
    (e) None of the above

    ANSWER:
    (a)

13. The point with polar coordinates, $r = 5$ and $\theta = 45°$ corresponds to which point in the complex plane?

    (a) $z = 5 + 45i$
    (b) $z = 5e^{(\pi/4)i}$
    (c) $z = 5 + \frac{\pi}{4}i$
    (d) $z = 5e^{45i}$
    (e) None of the above

    ANSWER:
    (b)

14. Which of the following complex numbers is not represented by a point on an axis of the complex plane?

    (a) $2i$
    (b) $i\sin\theta$
    (c) $(1 - i)(1 + i)$
    (d) e^{3i}
    (e) $e^{\pi i}$

    ANSWER:
    (d)

15. Order these complex numbers as nearest to farthest from the origin in the complex plane.

    (I) $2i$
    (I) $\sqrt{3}\cos(\pi/2) + i\sqrt{3}\sin(\pi/2)$
    (I) $1 - i$
    (I) $3e^{3i}$
    (I) e^{10i}

    (a) III, II, I V, IV
    (b) V, III, II, I, IV
    (c) III, I, II, IV, V

    ANSWER:
    (b)

Chapter Eight

ConcepTests and Answers and Comments for Section 8.1

1. If $f(x) = \sin x$ and $g(x) = x^2$, match the graphs in Figure 8.1 to the following function definitions
   (a) $y = f(g(x))$
   (b) $y = g(f(x))$
   (c) $y = f(x) + g(x)$
   (d) $y = f(x) \cdot g(x)$

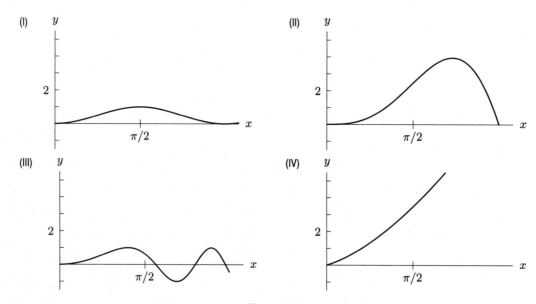

Figure 8.1

ANSWER:
   (I) (b)
   (II) (d)
   (III) (a)
   (IV) (c)

2. If $f(x) = \cos x$ and $g(x) = 2x - 1$, match the graphs in Figure 8.2 to the following function definitions

    (a) $y = f(g(x))$

    (b) $y = g(f(x))$

    (c) $y = f(x) + g(x)$

    (d) $y = f(x) \cdot g(x)$

(I)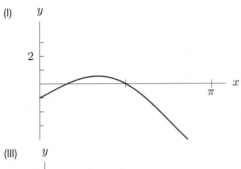

(II)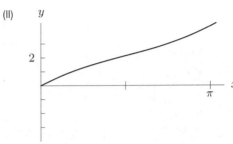

(III)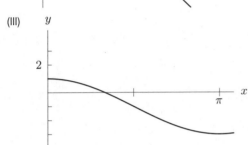

(IV)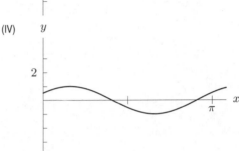

Figure 8.2

ANSWER:

  (I) (d)

  (II) (c)

  (III) (b)

  (IV) (a)

For Problems 3–6, identify the statement as (a) true, (b) false or (c) unknown.

3. For all functions f and g, $f(g(x)) = f(x) \cdot g(x)$.

    ANSWER:

    (b) False

4. For all functions f and g, $f(g(x)) = f(x) + g(x)$.

    ANSWER:

    (b) False

5. For all functions f and g, $f(g(x)) = g(f(x))$.

    ANSWER:

    (b) False

6. For some functions f and g, $f(g(x)) = g(f(x))$.

    ANSWER:

    True

7. If $f(t) = e^t$ and $g(t) = t^2$, then which of the following is $f(g(t))$?

    (a) $e^t \cdot t^2$

    (b) e^{2t}

    (c) e^{t^2}

    (d) et^2

    (e) None of the above

    ANSWER:

    (c)

8. If $f(t) = e^t$ and $g(t) = t^2$, then which of the following is $g(f(t))$?

   (a) $t^2 \cdot e^t$

   (b) e^{2t}

   (c) e^{t^2}

   (d) te^{2t}

   (e) None of the above

   ANSWER:

   (b)

9. If $f(x) = \sin x$ and $g(x) = 3x^2 + x$, then which of the following is $g(f(x))$?

   (a) $3\sin(x^2) + \sin x$

   (b) $\sin(3x^2 + x)$

   (c) $3\sin^2(x) + \sin x$

   (d) $3(\sin^2(x) + \sin x)$

   (e) None of the above

   ANSWER:

   (c)

10. If $f(x) = \sin x$ and $g(x) = 3x^2 + x$, then which of the following is $f(g(x))$?

    (a) $3\sin(x^2) + \sin x$

    (b) $\sin(3x^2 + x)$

    (c) $3\sin^2(x) + \sin x$

    (d) $\sin 3x^2 + x$

    (e) None of the above

    ANSWER:

    (b)

11. The cost of buying gasoline in New York depends upon the number of gallons pumped: $C = f(n)$. The conversion rate for dollars to Euros is $E = g(D)$. The conversion of gallons to liters is $G = h(L)$. Which of the following compositions creates a function for a European person to find the cost of New York gasoline in Euros per liter?

    (a) $h(f(g(x)))$

    (b) $f(g(h(x)))$

    (c) $g(f(h(x)))$

    (d) $g(h(f(x)))$

    (e) None of the above

    ANSWER:

    (c)

A ferris wheel operator uses a composite function to give the height above ground of riders from the o'clock positions. $f(x)$ gives the angle in degrees of the point on the ferris wheel at the x-o'clock position. $g(x)$ gives the height above the ground of the point on the ferris wheel corresponding to an angle of x degrees. The function $h(x) = g(f(x))$ is composed from Table 8.1 and Table 8.2.

For Problems 12–16, use the two tables to find the value of the given expression.

Table 8.1

x	1	2	3	4	5	...	12
$f(x)$	60°	30°	0°	330°	300°	...	90°

Table 8.2

x	0	30	60	90	120	150	...	330
$g(x)$	4	6	$4+2\sqrt{3}$	8	$4+2\sqrt{3}$	6	...	$4-2\sqrt{3}$

12. The value of $h(1)$?
    (a) 0   (b) 2   (c) 4   (d) 6   (e) $4+2\sqrt{3}$
    ANSWER:
    (e)

13. The value of $h(6)$?
    (a) 0   (b) 2   (c) 4   (d) 6   (e) $4+2\sqrt{3}$
    ANSWER:
    (a)

14. The value of $h(10)$?
    (a) 0   (b) 2   (c) 4   (d) 6   (e) $4+2\sqrt{3}$
    ANSWER:
    (d)

15. The value of $g(210)$?
    (a) 0   (b) 2   (c) 4   (d) 6   (e) $4+2\sqrt{3}$
    ANSWER:
    (b)

16. The value of $h(7)$?
    (a) 0   (b) 2   (c) 4   (d) $4+2\sqrt{3}$ (e) $4-2\sqrt{3}$
    ANSWER:
    (e)

ConcepTests and Answers and Comments for Section 8.2

For Problems 1–7, identify the statement as (a) true or (b) false.

1. Invertible for a function means that the function passes the Horizontal Line Test.
   ANSWER:
   (a) True

2. If $f(x) = \cos x$ then $f^{-1}(x)$ exists.
   ANSWER:
   (b) False
   COMMENT:
   Point out that $\cos^{-1} x$ is not an inverse for $\cos x$, but only for a restricted domain version of it.

3. If $f^{-1}(x) = \arcsin x$ then $f(x) = \sin x$.
   ANSWER:
   (a) True
   COMMENT:
   Strictly speaking, $f(x)$ is $\sin x$ restricted to the domain $-\pi/2 \le x \le \pi/2$.

4. If $f(x) = e^x$ then $f^{-1}(x) = \log x$.
   ANSWER:
   (b) False
   COMMENT:
   Ask the student what the inverse of e^x is. ($\ln x$).

5. If $f(x) = x^3$ then $f^{-1}(x) = \frac{1}{x^3}$.
   ANSWER:
   (b) False

6. If $f(x) = 3x$ then $f^{-1}(x) = \frac{1}{3}x$.
   ANSWER:
   (a) True

7. If $f(x) = \log x$ then $f^{-1}(x) = 10^x$.
   ANSWER:
   (a) True

8. The cost, in dollars, of buying gasoline in New York depends upon the number of gallons pumped: $C = f(n)$. The units of $f^{-1}(5)$ are
   (a) US dollars
   (b) US gallons
   (c) No units
   (d) None of the above

   ANSWER:
   (b)

9. If $g(x)$ is invertible if which of the following exists?
   (a) $g(x^{-1})$
   (b) $g^{-1}(x)$
   (c) $(g(x))^{-1}$
   (d) $\frac{1}{g(x)}$
   (e) None of the above

   ANSWER:
   (b)

For Problems 10–14, use Figure 8.3 to identify the statement as (a) true, (b) false or (c) unknown.

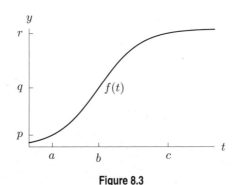

Figure 8.3

10. $f^{-1}(a) = p$

ANSWER:

(b) False

COMMENT:

The answer "unknown" could be defended by arguing that the scale is unknown and therefore this could be true as a special case, if $p = a$.

11. $f^{-1}(p) = a$

ANSWER:

(a) True

12. $f^{-1}(\frac{q+r}{2})$ exists.

ANSWER:

(a) True

13. $f^{-1}(0)$ does not exist.

ANSWER:

(b) False, we have to assume we are given the complete graph.

14. If $p < y < q$ then $a < f^{-1}(y) < b$.

ANSWER:

(a) True

15. Which of the following functions require a restricted domain for the inverse function to be defined?

(I) $y = |x|$

(II) $y = \sqrt{1 - x^2}$

(III) $y = \sqrt[3]{x}$

(IV) $y = \frac{1}{x}$

(V) $y = \sin x$

(VI) $y = e^x$

(VII) $y = \begin{cases} x & \text{if} & 0 < x < 5 \\ 5 & \text{if} & x \geq 5 \end{cases}$

(a) (II), (III), (IV)

(b) (I), (II), (III), (VII)

(c) (I), (II), (V)

(d) (I), (II), (V), (VII)

ANSWER:

(d)

COMMENT:

These are classic functions and you may want to require students to work without a graphing utility.

16. In Figure 8.4 which graphs are of invertible functions?

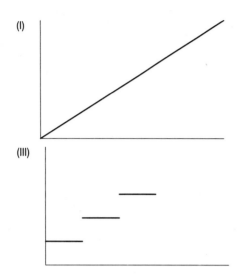

Figure 8.4

ANSWER:
(I) and (II)

17. In Figure 8.5 which graph shows the inverse of the function in (I)? The same scale is used on all axes.

   (a) The inverse does not exist
   (b) (II)
   (c) (III)
   (d) (IV)
   (e) None of the above

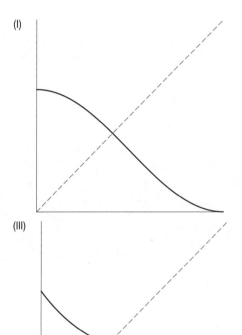

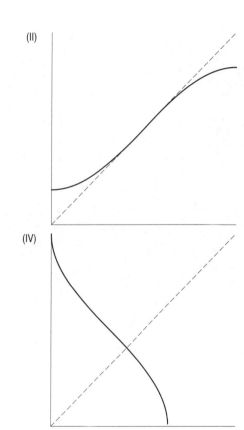

Figure 8.5

ANSWER:
(d)

ConcepTests and Answers and Comments for Section 8.3 ━━━━━

For Problems 1–3, use the graphs of the two functions $r(x)$ and $s(x)$ to match the graphs (I) to (IV).

1. Use Figure 8.6 to identify:

(a) $s(x) + r(x)$ (b) $s(x) - r(x)$ (c) $s(x) * r(x)$ (d) $s(x)/r(x)$

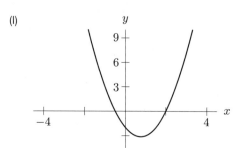
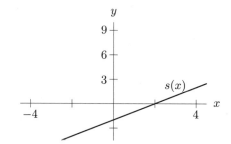

(I)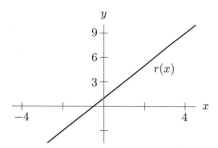

(II)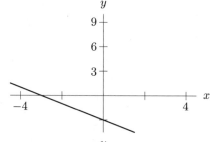

(III)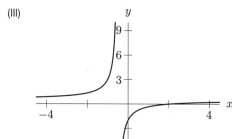

(IV)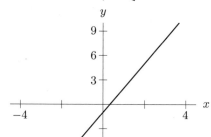

Figure 8.6

ANSWER:

(I) (c)
(II) (b)
(III) (d)
(IV) (a)

2. Use Figure 8.7 to identify:

(a) $r(x) + s(x)$ (b) $r(x) - s(x)$ (c) $r(x) * s(x)$ (d) $r(x)/s(x)$

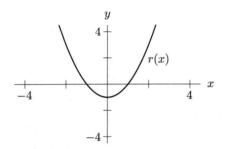

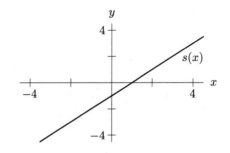

(I)

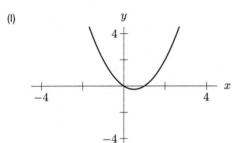

(II)

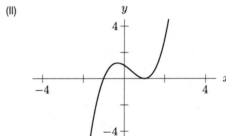

(III)

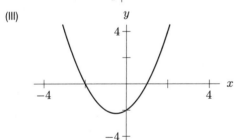

(IV)

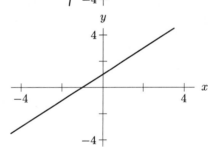

Figure 8.7

ANSWER:
(I) (b)
(II) (c)
(III) (a)
(IV) (d)

3. Use Figure 8.8 to identify:

(a) $r(x) + s(x)$ (b) $r(x) - s(x)$ (c) $r(x) * s(x)$ (d) $r(x)/s(x)$

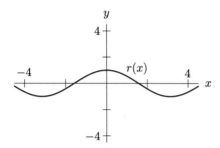

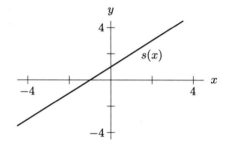

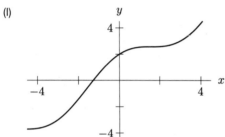

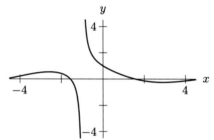

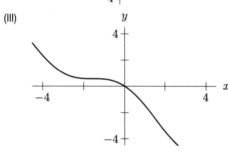

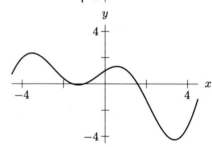

Figure 8.8

ANSWER:

(I) (a)
(II) (d)
(III) (b)
(IV) (c)

4. From which of the following function definitions is it easiest to identify the zeros?

(a) $(x - 2) \cdot (x + 3)$
(b) $e^x \cdot (5x^2 + 3x - 4)$
(c) $3x^2 \cdot (x^2 + x - 6)$
(d) $e^{2t} \cdot \sin(3t + 4)$
(e) All are equally easy

ANSWER:
(a)
COMMENT:
Discuss the order of difficulty. Note that this may be a matter of taste.

For Problems 5–7, find all the zeros of the given function by factoring.

5. $g(x) = x^2 - 5x$

   (a) 5
   (b) $\sqrt{5}$
   (c) 0
   (d) -5
   (e) None of the above

   (a) 5
   (b) $\sqrt{5}$
   (c) 0 and 5
   (d) 0 and $\sqrt{5}$

   ANSWER:
   (c)

6. $g(x) = e^{2x} - 2e^x$

   (a) 0
   (b) 2
   (c) $\ln 2$
   (d) e
   (e) None of the above

   ANSWER:
   (c)

7. $h(x) = xe^x + e^x$

   (a) -1
   (b) 0
   (c) 1
   (d) ± 1
   (e) None of the above

   ANSWER:
   (a)

8. Match the use of the product functions, (a) - (d), to the applications (I) - (IV).
   (I) oscillation with increasing amplitude
   (II) Calculating volume
   (III) Calculating area
   (IV) Damped oscillation

   (a) $(x + 2) \cdot (x + 5)$
   (b) $e^{-x} \cdot \sin(3x + \pi/4)$
   (c) $x^2 \cdot (x + 1)$
   (d) $(2x + 1) \cdot \cos(0.5x)$

   ANSWER:

   (a) (III)
   (b) (IV)
   (c) (II)
   (d) (I)

9. If the graphs of two functions, $r(t)$ and $s(t)$, intersect at the point $(3, 5)$, order the following values from least to greatest.

   (a) $r(3) + s(3)$
   (b) $r(3) - s(3)$
   (c) $r(3) * s(3)$
   (d) $r(3)/s(3)$

   ANSWER:
   (b), (d), (a), (c)

10. If $r(t) = s(t)$ at the point $t = a$, match the following combination function values, (a) - (d), to the values, (I) - (IV).

   (I) 0

   (II) 1

   (III) $2r(a)$

   (IV) $r(a)^2$

   (a) $r(a) + s(a)$

   (b) $r(a) - s(a)$

   (c) $r(a) * s(a)$

   (d) $r(a)/s(a)$

   ANSWER:

   (I) (b)

   (II) (d)

   (III) (a)

   (IV) (c)

   COMMENT:

   What if $r(a) = 0$?

Chapter Nine

ConcepTests and Answers and Comments for Section 9.1 ——————

For Problems 1–3, identify the relationship as
(a) direct proportionality   (b) indirect proportionality   (c) not a proportion

1. Fahrenheit temperature as a function of its Celsius temperature.
   ANSWER:
   (c)

2. The exchange rate of the US dollar as a function of the Euro value.
   ANSWER:
   (a)

3. Velocity as a function of the time traveled a given distance.
   ANSWER:
   (b)

For Problems 4–15, consider each function to be in the form $y = k \cdot x^p$, and identify k or p as requested. Answer with (e) if the function is not a power function.

4. If $y = \sqrt{2x}$, give p
   (a) $\sqrt{2}$
   (b) 2
   (c) 1
   (d) $\frac{1}{2}$
   (e) Not a power function
   ANSWER:
   (d)

5. If $y = \sqrt{2x}$, give k
   (a) $\sqrt{2}$
   (b) 2
   (c) 1
   (d) $\frac{1}{2}$
   (e) Not a power function
   ANSWER:
   (a)

6. If $A = \pi r^2$, give p
   (a) π
   (b) 2
   (c) 1
   (d) $\frac{1}{2}$
   (e) Not a power function
   ANSWER:
   (b)

7. If $A = \pi r^2$, give k.
   (a) π
   (b) 2
   (c) 1
   (d) $\frac{1}{2}$
   (e) Not a power function
   ANSWER:
   (a)

8. If $y = x^2 + 1$, give k.

   (a) 0
   (b) $\frac{1}{2}$
   (c) 1
   (d) 2
   (e) Not a power function

   ANSWER:

   (e)

9. If $y = \dfrac{1}{\pi x}$, give k.

   (a) -1
   (b) $\frac{1}{\pi}$
   (c) 1
   (d) $-\pi$
   (e) Not a power function

   ANSWER:

   (b)

10. If $y = \dfrac{1}{\pi x}$, give p.

    (a) -1
    (b) $\frac{1}{\pi}$
    (c) 1
    (d) $-\pi$
    (e) Not a power function

    ANSWER:

    (a)

11. If $y = x^{\sqrt{2}}$, give p.

    (a) 0
    (b) $\sqrt{2}$
    (c) $\frac{1}{2}$
    (d) 1
    (e) Not a power function

    ANSWER:

    (b)

12. If $y = x^{\sqrt{2}}$, give k.

    (a) 0
    (b) $\sqrt{2}$
    (c) $\frac{1}{2}$
    (d) 1
    (e) Not a power function

    ANSWER:

    (d)

13. If $y = x^2 + x$, give k.

    (a) 0
    (b) 1
    (c) 2
    (d) -1
    (e) Not a power function

    ANSWER:

    (e)

14. If $y = -3$, give k.

   (a) 0
   (b) 1
   (c) 2
   (d) -3
   (e) Not a power function

   ANSWER:
   (d)

15. If $y = -3$, give p.

   (a) 0
   (b) 1
   (c) 2
   (d) -3
   (e) Not a power function

   ANSWER:
   (a)

16. The value of x^5 is greater than the value of x^3 for

   (a) All reals
   (b) All positive reals
   (c) $x > 1$ and $-1 < x < 0$
   (d) $|x| > 1$
   (e) None of the above

   ANSWER:
   (c)

17. The value of x^4 is greater than the value of x^2 for

   (a) All reals
   (b) All positive reals
   (c) $x > 1$
   (d) $|x| > 1$
   (e) None of the above

   ANSWER:
   (d)

18. The value of x^{-5} is less than the value of x^{-3} for

   (a) All reals
   (b) All positive reals
   (c) $x > 1$
   (d) $|x| > 1$
   (e) None of the above

   ANSWER:
   (c)

   COMMENT:
   Ask the students to solve the inequality $x^{-5} - x^{-3} < 0$.

19. The value of x^{-4} is less than the value of x^{-2} for

   (a) All reals
   (b) All positive reals
   (c) $x > 1$
   (d) $|x| > 1$
   (e) None of the above

   ANSWER:
   (d)

ConcepTests and Answers and Comments for Section 9.2

1. For each of the following functions answer (a) yes if it is a polynomial function, or (b) no, if not.

   (I) $y = 1 + x$
   (II) $y = 1/x$
   (III) $y = 6$
   (IV) $y = \sqrt{2x}$
   (V) $y = \sqrt{2}x$
   (VI) $y = \pi r^2$
   (VII) $y = \sin(x^2 - x + 6)$
   (VIII) $y = \sin^2 x + 2\sin x + 1$
   (IX) $y = (1 + x)^2$
   (X) $y = 10^x$
   (XI) $y = f(t) = \dfrac{s}{t}$
   (XII) $y = f(w) = 2l + 2w$

   ANSWER:

   (I)   (a) yes
   (II) (b) no
   (III) (a) yes
   (IV) (b) no
   (V)   (a) yes
   (VI) (a) yes
   (VII) (b) no
   (VIII) (b) no
   (IX) (a) yes
   (X)   (b) no
   (XI) (b) no
   (XII) (a) yes

For Problems 2–3, use Figure 9.1 which shows both the long-run and short-run behavior of the functions. For each function (I) to (VIII), identify the requested number.

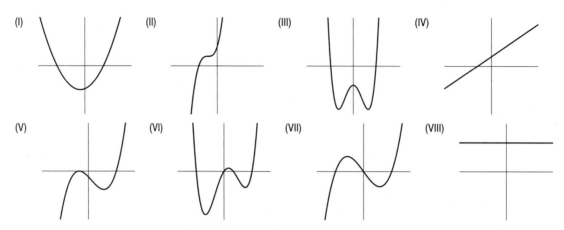

Figure 9.1

2. Identify the number of distinct zeros of each function. Use (a) for no zeros, (b) for 1 zero, (c) for 2 zeros, (d) for 3 zeros and (e) for 4 zeros.

    ANSWER:

    (I) (c) 2
   (II) (b) 1
  (III) (c) 2
   (IV) (b) 1
    (V) (c) 2
   (VI) (e) 4
  (VII) (d) 3
 (VIII) (a) 0

3. If each polynomial function shown has a degree from (a) 0 to (e) 4, identify the degree of each function from its graph.

    ANSWER:

    (I) (c) 2
   (II) (d) 3
  (III) (e) 4
   (IV) (b) 1
    (V) (d) 3
   (VI) (e) 4
  (VII) (d) 3
 (VIII) (a) 0

    COMMENT:

    Ask the students why the degree in part (I) could not be 4. The reason is that the curve is not flat enough near its vertex.

For Problems 4–9, identify the statement as being true
(a) Always   (b) Sometimes   (c) Never

4. Polynomial functions are power functions
   ANSWER:
   (b) Sometimes

5. Power functions are polynomial functions.
   ANSWER:
   (a) Always

6. Linear functions are polynomial functions.
   ANSWER:
   (a) Always

7. If the leading term is listed first, the polynomial is in standard form.
   ANSWER:
   (b) Sometimes

8. If the degree of the polynomial is 3, it can have 4 zeros.
   ANSWER:
   (c) Never

9. If $p(x) = a + 2x + bx^3$, then 2 is a coefficient.
   ANSWER:
   (a) Always

ConcepTests and Answers and Comments for Section 9.3 ▬▬▬▬▬

For Problems 1–8, the long-run behavior of each function is shown in the graph. Identify its possible equation.

1. In Figure 9.2
   (a) $y = (x - 1)^2(x + 2)$
   (b) $y = -(x - 1)^2(x + 2)$
   (c) $y = -(x + 1)^2(x - 2)$
   (d) $y = (x + 1)^2(x - 2)$
   (e) None of the above

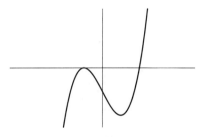

Figure 9.2

   ANSWER:
   (d)

2. In Figure 9.3

    (a) $y = x(x^2 - 9)(x - 1)$

    (b) $y = (x^2 - 1)(x^2 - 9)$

    (c) $y = (x - 3)(x + 3)(x - 1)$

    (d) $y = x^4 - 4$

    (e) None of the above

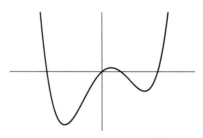

Figure 9.3

ANSWER:

(a)

3. In Figure 9.4

    (a) $y = (x + 2)^3$

    (b) $y = -(x + 2)^3$

    (c) $y = (x - 2)^3$

    (d) $y = -(x - 2)^3$

    (e) None of the above

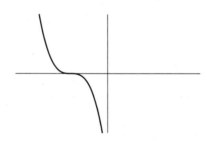

Figure 9.4

ANSWER:

(b)

4. In Figure 9.5

   (a) $y = (x + 2)(x - 3)$
   (b) $y = -x^2 - 6$
   (c) $y = (x - 2)(x + 3)$
   (d) $y = x(x - 2)(x + 3)$
   (e) None of the above

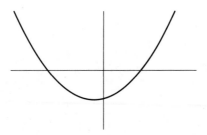

Figure 9.5

ANSWER:
(c)

5. In Figure 9.6

   (a) $y = (x - 1)(x + 2)$
   (b) $y = -(x + 1)(x - 2)$
   (c) $y = -(x - 1)(x + 2)$
   (d) $y = (x - 1)(x - 2)$
   (e) None of the above

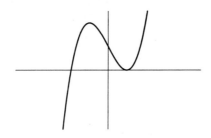

Figure 9.6

ANSWER:
(e)

6. In Figure 9.7

    (a) $y = 3 - (x + 1)^2$

    (b) $y = -x^2(x + 1)^2$

    (c) $y = 3 - x^2(x + 1)^2$

    (d) $y = x^2(x + 1)^2 + 3$

    (e) None of the above

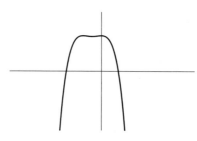

Figure 9.7

ANSWER:

(c)

7. In Figure 9.8, with a and b positive.

    (a) $y = (x + a)(x - b)$

    (b) $y = (x - a)(x + b)$

    (c) $y = x(x - a)(x - b)$

    (d) $y = x(x + a)(x - b)$

    (e) None of the above

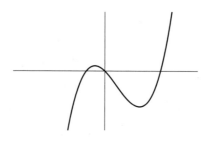

Figure 9.8

ANSWER:

(d)

8. In Figure 9.9, the function $y = (x+1)^4 - 3$ is graphed as a dashed line. Identify the equation of the solid graph.

   (a) $y = (x-1)(x+3)$

   (b) $y = (x+1)^2 - 3$

   (c) $y = (x+1)^3 - 3$

   (d) $y = (x+1)^6 - 3$

   (e) None of the above

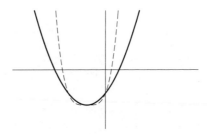

Figure 9.9

ANSWER:

(b)

COMMENT:

Ask for reasons that (a) can be eliminated, such as its value at $x = 0$ is not the same as the given functions.

For Problems 9–15, the polynomial $f(x)$ has exactly three different zeros, identify each statement as being true
(a) Always   (b) Sometimes   (c) Never

9. The function $f(x)$ is an odd function.

   ANSWER:

   (a) Sometimes

10. The function $f(x)$ is an even function.

   ANSWER:

   (c) Sometimes

11. The function $f(x)$ is a third degree polynomial.

   ANSWER:

   (b) Sometimes

12. The function $f(x)$ is invertible.

   ANSWER:

   (c) Never

13. The function has multiple zeros

   ANSWER:

   (b) Sometimes

14. The polynomial has degree 5.

   ANSWER:

   (b) Sometimes

15. The polynomial has degree four.

   ANSWER:

   (c) Sometimes

ConcepTests and Answers and Comments for Section 9.4

1. For each of the following functions answer (a) yes if it is a rational function, and (b) no, if not.

    (I) $y = 1/x + 1$

    (II) $y = 1 + x$

    (III) $y = 6$

    (IV) $y = \sqrt{\dfrac{x+1}{x^2-3}}$

    (V) $y = \dfrac{\sqrt{2}x}{x-1}$

    (VI) $y = \sin(\dfrac{x^2}{-x+6})$

    (VII) $y = \dfrac{\sin^2 x}{x}$

    (VIII) $y = (1+x)^{-2}$

    (IX) $y = \dfrac{10^x}{x+10}$

    (X) $y = \dfrac{(x+1)(x-3)}{x+1}$

    (XI) $y = \dfrac{\ln x}{x}$

    ANSWER:

    (I) (a) yes, can be written as $(x+1)/x$

    (II) (a) yes, can be written $y = \frac{1+x}{1}$

    (III) (a) yes, can be written $y = \frac{6}{1}$

    (IV) (b) no

    (V) (a) yes

    (VI) (b) no

    (VII) (b) no

    (VIII) (a) yes

    (IX) (b) no

    (X) (a) yes

    (XI) (b) no

2. The long-run behavior of $r(t) = \dfrac{2t+1}{1+\frac{1}{2}t}$ is

    (a) $y = t$

    (b) $y = 2$

    (c) $y = 4$

    (d) $y = 2t$

    (e) None of the above

    ANSWER:

    (c)

3. The long-run behavior of $r(t) = \dfrac{t^3+1}{1+\frac{1}{2}t}$ is

    (a) $y = t^3$

    (b) $y = 2t^2$

    (c) $y = 2t$

    (d) $y = t^2$

    (e) None of the above

    ANSWER:

    (b)

4. The long-run behavior of $r(t) = \dfrac{1 + 2t}{t^2 + \frac{1}{3}}$ is

(a) $y = 2/t$
(b) $y = 2/t^2$
(c) $y = 6t$
(d) $y = 2$
(e) None of the above

ANSWER:

(a)

For Problems 5–9, identify the statement as being true

(a) Always   (b) Sometimes   (c) Never

5. A rational function has a vertical asymptote.
ANSWER:
(b) Sometimes

6. The domain of a rational function does not include values where the denominator is zero.
ANSWER:
(a) Always

7. A rational function has a horizontal asymptote.
ANSWER:
(b) Sometimes

8. The range of a rational function is all real numbers.
ANSWER:
(b) Sometimes

9. The domain of a rational function is all real numbers.
ANSWER:
(b) Sometimes

10. If $Q(t)$ is the amount of food supply at time t, and $P(t)$ is the population at time t, then which of the following functions represents food per capita?

(a) $Q(t) + P(t)$
(b) $Q(t) * P(t)$
(c) $Q(t)/P(t)$
(d) $P(t)/Q(t)$
(e) None of the above

ANSWER:

(c)

11. If $C(n)$ is the cost to make n items, then the function $f(n) = C(n)/n$ is ?

(a) The average rate of change of $C(n)$
(b) The average cost of an item
(c) Cost per capita
(d) A constant, C
(e) None of the above

ANSWER:

(b)

12. Match long-run behaviors, (a)-(d), as $x \to \infty$ to the graphs (I) to (IV) in Figure 9.10

    (a) $y \to -\infty$

    (b) $y = -2$

    (c) $y = 0$

    (d) $y \to \infty$

(I)
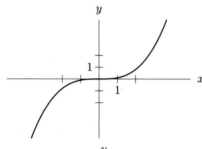

(II)
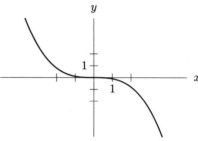

(III)
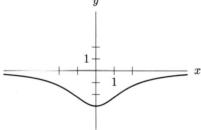

(IV)
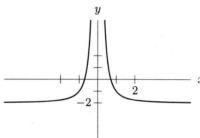

Figure 9.10

    ANSWER:

  (I) (d)

  (II) (a)

  (III) (c)

  (IV) (b)

ConcepTests and Answers and Comments for Section 9.5

For Problems 1–6, identify the statement as being true

    (a) Always   (b) Sometimes   (c) Never

1. If the numerator of a rational function is a non-zero constant, then the function has no zeros.
    ANSWER:
    (a) Always

2. A rational function has both a vertical and horizontal asymptote.
    ANSWER:
    (b) Sometimes

3. If the denominator of a rational function is always non-zero, then the function's graph does not have a hole.
    ANSWER:
    (a) Always

4. A function approaches values of its horizontal asymptote, but never equals that value.
    ANSWER:
    (b) Sometimes

5. The zeros of a rational function are where the denominator equals zero.
    ANSWER:
    (c) Never

6. There can be two vertical intercepts of a rational function.
    ANSWER:
    (c) Never

7. Match the graphs (I) to (IV) in Figure 9.11, to the equation $y = (x - a)(x - b)/(x - c)$ with the following conditions

    (a) $a = b$

    (b) $a = c$

    (c) $a < b < c$

    (d) $a < c < b$

(I)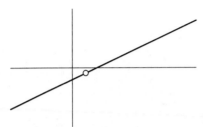

(II)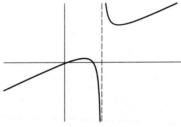

(III)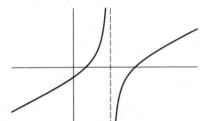

(IV)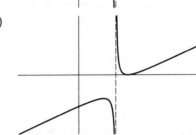

Figure 9.11

ANSWER:

  (I) (b)

  (II) (c)

 (III) (d)

 (IV) (a)

8. Which of the four tables corresponds to the equation $y = \dfrac{1}{(x - 3)(x - 2)}$? Do not use a calculator.

(a)

Table 9.1

x	2.7	2.8	2.9	3	3.1	3.2	3.3
y	−4.77	−6.25	−11.11	0	9.09	4.17	2.56

(b)

Table 9.2

x	2.7	2.8	2.9	3	3.1	3.2	3.3
y	4.77	6.25	11.11	0	9.09	4.17	2.56

(c)

Table 9.3

x	2.7	2.8	2.9	3	3.1	3.2	3.3
y	−4.77	−6.25	−11.11	Error	9.09	4.17	2.56

(d)

Table 9.4

x	2.7	2.8	2.9	3	3.1	3.2	3.3
y	4.77	6.25	11.11	Error	9.09	4.17	2.56

ANSWER:

  (c)

9. Which equation matches the graph in Figure 9.12.

    (a) $y = 1/(x^2 - 4)$
    (b) $y = (x^2 - 4)/(2 - x^2)$
    (c) $y = (2 - x^2)/(x^2 - 4)$
    (d) $y = (x^2 - 2)/(x^2 - 4)$

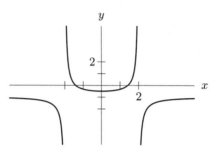

Figure 9.12

ANSWER:
(c)

For Problems 10–16, a rational function is given. Find the number of vertical asymptotes in its graph

    (a) 0   (b) 1   (c) 2   (d) 3   (e) 4 or more

10. ANSWER:

11. $r(x) = \dfrac{(x+2)(x+1)}{(x+2)}$
    ANSWER:
    (a)

12. $r(x) = \dfrac{x+2}{(x^2-1)}$
    ANSWER:
    (c)

13. $r(x) = \dfrac{(x+5)^3}{(x^2-5)}$
    ANSWER:
    (c)

14. $r(x) = \dfrac{x+4}{(x^2+x)}$
    ANSWER:
    (c)

15. $r(x) = \dfrac{3x-4}{x^2+1}$
    ANSWER:
    (a)

16. $r(x) = \dfrac{1}{x^5}$
    ANSWER:
    (b)

ConcepTests and Answers and Comments for Section 9.6

For Problems 1–7, identify the statement as (a) true or (b) false.

1. A positive increasing exponential function $f(x)$ eventually grows larger than any power function as $x \to \infty$.
   ANSWER:
   (a) True

2. A positive decreasing exponential function $g(x)$ eventually approaches zero faster than any positive decreasing power function as $x \to \infty$.
   ANSWER:
   (a) True

3. The value $f(x) = x^4$ is less than $g(x) = 4^x$, for all x.
   ANSWER:
   (b) False

4. The values of $f(x) = x^4$ are greater than $g(x) = 4^x$ as $x \to \infty$.
   ANSWER:
   (b) False

5. For $x > 0$, the graph of a power function, $y = x^p$ is concave up, for all p.
   ANSWER:
   (b) False

6. The graph of an exponential function, $y = b^x$ is concave up, for all $b > 0$, $b \neq 1$.
   ANSWER:
   (a) True

7. The value of $f(x) = \dfrac{\log x}{x}$ approaches zero as $x \to \infty$.
   ANSWER:
   (a) True

8. Without using a calculator, order the values from least to greatest.

   (a) 20^2
   (b) 2^{20}
   (c) 30^3
   (d) 3^{30}
   (e) 1^{500}

   (a) $1^{500}, 20^2, 30^3, 2^{20}, 3^{30}$
   (b) $20^2, 30^3, 2^{20}, 3^{30}, 1^{500}$
   (c) $1^{500}, 2^{20}, 3^{30}, 20^2, 30^3$

   ANSWER:
   (a)

9. Without using a calculator, order the values from least to greatest.

   (a) 20^{-2}
   (b) 2^{-20}
   (c) 30^{-3}
   (d) 3^{-30}
   (e) 1^{-500}

   (a) $1^{-500}, 20^{-2}, 30^{-3}, 2^{-20}, 3^{-30}$
   (b) $3^{-30}, 2^{-20}, 30^{-3}, 20^{-2}, 1^{-500}$
   (c) $1^{-500}, 2^{-20}, 3^{-30}, 20^{-2}, 30^{-3}$

   ANSWER:
   (b)
   COMMENT:
   Since the numbers in this problem are exactly the reciprocals of the numbers in the previous problem, the order is reversed.

10. What is the long-run behavior of $y = f(x) = \dfrac{x^2}{2^x}$?

(a) $y \to \infty$
(b) $y = 2$
(c) $y = 1$
(d) $y \to 0$
(e) None of the above

ANSWER:

(d)

ConcepTests and Answers and Comments for Section 9.7

1. The equations (I) to (VII) are listed on a graphing calculator as regression equations. Match them to the names of the regression types. (a) Power   (b) Exponential   (c) Quadratic   (d) Linear
(e) Cubic   (f) Logistic   (g) Sinusoidal

(I) $y = b + ax$
(II) $y = ab^\wedge x$
(III) $y = ax^\wedge b$
(IV) $y = ax^\wedge 3 + bx^\wedge 2 + cx + d$
(V) $y = a \cdot \sin(bx + c) + d$
(VI) $y = ax^\wedge 2 + bx + c$
(VII) $y = c/(1 + ae^\wedge(-bx))$

ANSWER:

(I) (d)
(II) (b)
(III) (a)
(IV) (e)
(V) (g)
(VI) (c)
(VII) (d)

2. Which graph in Figure 9.13, corresponds to the data in Table 9.5

Table 9.5

t	1	2	3	4	5	6	7	8	9
$f(t)$	1.9	3.6	5.1	6.4	7.5	8.4	9.1	9.6	9.9

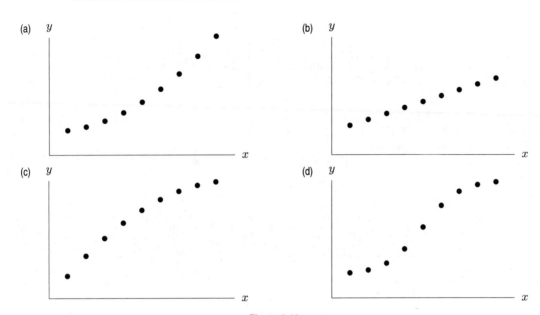

Figure 9.13

ANSWER:

(c)

3. Which of the four tables has data corresponding to the graph in Figure 9.14?

(a)

Table 9.6

t	1	2	3	4	5	6	7	8	9
$f(t)$	2.1	2.4	3	4.2	6	7.9	9	9.6	9.9

(b)

Table 9.7

t	1	2	3	4	5	6	7	8	9
$f(t)$	2.5	3	3.5	4	4.5	5	5.5	6	6.5

(c)

Table 9.8

t	1	2	3	4	5	6	7	8	9
$f(t)$	1.9	3.6	5.1	6.4	7.5	8.4	9.1	9.6	9.9

(d)

Table 9.9

t	1	2	3	4	5	6	7	8	9
$f(t)$	2.1	2.4	2.9	3.6	4.5	5.6	6.9	8.4	10.1

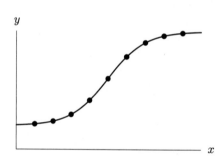

Figure 9.14

ANSWER:
(a)

4. Which of the regression types is well suited to represent the data shown in Figure 9.15?

    (a) Linear

    (b) Power

    (c) Exponential

    (d) None of the above

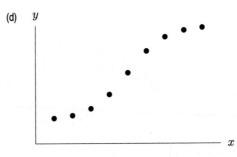

Figure 9.15

ANSWER:

(d)

COMMENT:

It appears to be a logistic equation.

For Problems 5–8, identify the statement as (a) true or (b) false.

5. The best-fit exponential equation is $y = ab^x + c$, where $c \neq 0$.

    ANSWER:

    (b) False

6. The graph of the best-fit power equation goes through the origin.

    ANSWER:

    (a) True

7. An exponential best-fit equation can be written either as $y = ab^x$ or $y = ae^{kx}$.

    ANSWER:

    (a) True

8. If a and b have positive values and $b \neq 1$ then the graph of $y = ab^x$ is concave up.

    ANSWER:

    (a) True

Chapter Ten

ConcepTests and Answers and Comments for Section 10.1 ━━━━━━━━

For Problems 1–11, identify the statement as being true
(a) Always   (b) Sometimes   (c) Never

1. The difference of two vectors depends on the order of subtraction.
   ANSWER:
   (b) Sometimes

2. The sum of two vectors depends on the order of addition.
   ANSWER:
   (c) Never

3. Scalars can have negative values.
   ANSWER:
   (b) Sometimes

4. The zero vector has no direction.
   ANSWER:
   (a) Always

5. $\|\vec{u}\| + \|\vec{v}\| = \|\vec{u} + \vec{v}\|$
   ANSWER:
   (b) Sometimes

6. $\|\vec{u}\| + \|\vec{u}\| = 2\|\vec{u}\|$
   ANSWER:
   (a) Always

7. $(a + b)\vec{u} = a\vec{u} + b\vec{u}$
   ANSWER:
   (a) Always

8. $\vec{v} + \vec{u} = \vec{u} + \vec{v}$
   ANSWER:
   (a) Always

9. $\|\vec{v}\|$ is non-negative
   ANSWER:
   (a) Always

10. $\vec{1}$ is the unit vector
    ANSWER:
    (c) Never

11. $\vec{0}$ is the zero vector
    ANSWER:
    (a) Always

12. Which of the following is a valid vector notation?

    (a) $|\vec{x}|$
    (b) $\|u\|$
    (c) $\overrightarrow{(3v)}$
    (d) $\vec{x}\,\vec{y}$
    (e) None of the above

       ANSWER:
       (e)

13. A vector $\vec{v}$ has magnitude 5 and direction due north. Identify each expression as being true or false.

(a) $\| - \vec{v} \| = -5$

(b) $\|3\vec{v}\| = 15$

(c) $3\|\vec{v}\| = 15$

(d) $-\vec{v}$ is a vector with direction due south

ANSWER:

(a) False

(b) True

(c) True

(d) True

ConcepTests and Answers and Comments for Section 10.2 ———————

For all problems in this section, $\vec{i}$ and $\vec{j}$ are the standard unit vectors and θ is the angle of a vector measured counterclockwise from the positive x-axis.

1. The value of $\|3\vec{i} + 2\vec{j}\|$ is

(a) 5

(b) $\sqrt{5}$

(c) 13

(d) $\sqrt{13}$

(e) None of the above

ANSWER:

(d)

2. The value of $\|3\vec{i} - 2\vec{j}\|$ is

(a) -13

(b) $-\sqrt{13}$

(c) 13

(d) $\sqrt{13}$

(e) None of the above

ANSWER:

(d)

3. The value of θ for $5\vec{i} + 12\vec{j}$ is

(a) $\cos^{-1}(12/5)$

(b) $\cos^{-1}(5/12)$

(c) $\sin^{-1}(12/13)$

(d) $\sin^{-1}(13/12)$

(e) None of the above

ANSWER:

(c)

4. The value of θ for $5\vec{i} - 12\vec{j}$ is

(a) $-\cos^{-1}(5/13)$

(b) $\cos^{-1}(5/13)$

(c) $-\sin^{-1}(5/13)$

(d) $\sin^{-1}(5/13)$

(e) None of the above

ANSWER:

(a)

5. The displacement vector from the point $(2, 2)$ to $(-3, 0)$ is

   (a) $-5\vec{i} + 2\vec{j}$
   (b) $5\vec{i} + 2\vec{j}$
   (c) $-5\vec{i} - 2\vec{j}$
   (d) $5\vec{i} - 2\vec{j}$
   (e) None of the above

   ANSWER:
   (c)

6. The displacement vector from the point $(-3, 0)$ to $(2, 2)$ is

   (a) $-5\vec{i} + 2\vec{j}$
   (b) $5\vec{i} + 2\vec{j}$
   (c) $-5\vec{i} - 2\vec{j}$
   (d) $5\vec{i} - 2\vec{j}$
   (e) None of the above

   ANSWER:
   (b)

7. A vector in the opposite direction to $-5\vec{i} + 2\vec{j}$ is

   (a) $5\vec{i} + 2\vec{j}$
   (b) $2\vec{i} - 5\vec{j}$
   (c) $-15\vec{i} + 6\vec{j}$
   (d) $5\vec{i}$
   (e) None of the above

   ANSWER:
   (e)

8. A vector of the same magnitude as $-5\vec{i} + 2\vec{j}$ is

   (a) $-15\vec{i} + 6\vec{j}$
   (b) $15\vec{i} + 6\vec{j}$
   (c) $\sqrt{29}\,\vec{j}$
   (d) $7\vec{i}$
   (e) None of the above

   ANSWER:
   (c)

9. Which vector does not have the same magnitude as the others?

   (a) $\vec{i} - \vec{j}$
   (b) $-\vec{i} + \vec{j}$
   (c) $\sqrt{2}\,\vec{j}$
   (d) $-\sqrt{2}\,\vec{i}$
   (e) All are the same magnitude

   ANSWER:
   (e)

10. Which vector does not have the same direction as the others?

    (a) $\vec{i} - \vec{j}$
    (b) $3\vec{i} - 3\vec{j}$
    (c) $\sqrt{2}\,\vec{i} - \sqrt{2}\,\vec{j}$
    (d) $\sqrt{2}\,\vec{j} - \sqrt{2}\,\vec{i}$
    (e) All are the same direction

    ANSWER:
    (d)

ConcepTests and Answers and Comments for Section 10.3 ━━━━━

1. If a vector $\vec{V}_{2005}$ shows oil prices at the end of each month in 2005, then the dimension of this vector is

   (a) 1
   (b) 7
   (c) 12
   (d) 365
   (e) None of the above

   ANSWER:
   (c)

2. A vector $\vec{V}_{2005}$ shows oil prices at the end of each month in 2005. Which of the following vectors adjusts the prices downward to compensate for 2% inflation?

   (a) $1.02\vec{V}_{2005}$
   (b) $\frac{1}{1.02}\vec{V}_{2005}$
   (c) $0.98\vec{V}_{2005}$
   (d) $\vec{V}_{2005} - 0.02$
   (e) None of the above

   ANSWER:
   (b)

3. If a vector starts at $(-1, 2)$ and ends at $(2, -1)$ then its displacement vector is

   (a) $\vec{i} + \vec{j}$
   (b) $3\vec{i} + 3\vec{j}$
   (c) $-3\vec{i} + 3\vec{j}$
   (d) $3\vec{i} - 3\vec{j}$
   (e) None of the above

   ANSWER:
   (d)

4. A computer user clicks on a screen pixel at $(-1, 1)$ and drags it to $(-3, 2)$. Using the same displacement vector, the point $(2, 1)$ moves to which of the following points?

   (a) $(1, 2)$
   (b) $(1, 1)$
   (c) $(0, 2)$
   (d) $(-4, 3)$
   (e) None of the above

   ANSWER:
   (c)

For Problems 5–10, consider the following situation. An exam has three parts: True/False worth 25%, Multiple Choice worth 25%, and Essay worth 50%. Your scores on the three parts of exam 1 are recorded in an exam vector $\vec{T}_1 = (t_1, m_1, e_1)$. Your second exam is $\vec{T}_2$ and $\vec{T}_1 = 1.2\vec{T}_2$.

5. True or False? The total score of exam 2 is greater than exam 1.

    ANSWER:

    False

6. True or False? Each section score has changed by the same amount.

    ANSWER:

    False

7. True or False? Each section score has changed by the same percent.

    ANSWER:

    True

8. Always, Sometimes, Never? Your Multiple Choice score could have been the same on the first two exams.

    ANSWER:

    Sometimes, but only when the Multiple Choice score is zero.

9. What is the average score of the two exams?

    (a) $0.5(\vec{T}_1 + \vec{T}_2)$
    (b) $\frac{1}{2}\vec{T}_1 + \frac{1}{2}\vec{T}_2$
    (c) $(\dfrac{t_1 + t_2}{2}, \dfrac{m_1 + m_2}{2}, \dfrac{e_1 + e_2}{2})$
    (d) All of the above
    (e) None of the above

    ANSWER:

    (e)

    COMMENT:

    The three expressions give the "average" vector but not a total score.

10. If $t_3 + m_3 + e_3 = t_4 + m_4 + e_4$ then $\vec{T}_3 = \vec{T}_4$

    (a) Always
    (b) Sometimes
    (c) Never

    ANSWER:

    (b)

ConcepTests and Answers and Comments for Section 10.4 ────

For Problems 1–10, identify the expression as being a
(a) Scalar   (b) Vector   (c) Undefined expression

1. $2\vec{v}$
     ANSWER:
     (b)

2. $\vec{v}^{\,2}$
     ANSWER:
     (c)

3. $\vec{v} \cdot \vec{v}$
     ANSWER:
     (a)

4. $\vec{v} + 2$
     ANSWER:
     (c)

5. $0.08 \cdot 1.16$
     ANSWER:
     (a)

6. $\vec{u} \cdot \vec{v} - \vec{u} \cdot \vec{w}$
     ANSWER:
     (a)

7. $\vec{u} \cdot (\vec{v} - \vec{w})$
     ANSWER:
     (a)

8. $\|\vec{u}\| \cdot \|\vec{v}\|$
     ANSWER:
     (a)

9. $\dfrac{1}{\|\vec{u}\|} \cdot \vec{u}$
     ANSWER:
     (b)

10. $\|\vec{u} \cdot \vec{v}\|$
     ANSWER:
     (c)

For Problems 11–17, $\vec{i}$ and $\vec{j}$ are the standard unit vectors and $\vec{u} = 3\vec{i} - 2\vec{j}$.

11. The value of $\vec{i} \cdot \vec{j}$ is

    (a) 0
    (b) 1
    (c) -1
    (d) $\sqrt{2}$
    (e) Not defined

    ANSWER:
    (a)

12. The value of $\vec{i} \cdot \vec{u}$ is

    (a) 0
    (b) 1
    (c) 3
    (d) $3\vec{i}$
    (e) $\sqrt{13}$
    (f) Not defined

    ANSWER:
    (c)

13. The value of $\vec{u} \cdot \vec{j}$ is

    (a) -2
    (b) 2
    (c) 3
    (d) $\sqrt{13}$
    (e) Not defined

    ANSWER:
    (a)

14. The value of $\vec{u} \cdot \vec{u}$ is

    (a) 1
    (b) 5
    (c) 13
    (d) $\sqrt{13}$
    (e) Not defined

    ANSWER:
    (c)

15. The value of $\|\vec{u}\|$ is

    (a) 0
    (b) 1
    (c) 13
    (d) $\sqrt{13}$
    (e) Not defined

    ANSWER:
    (d)

16. The value of $\vec{j} \cdot \vec{j}$ is

    (a) 0
    (b) 1
    (c) -1
    (d) $\sqrt{2}$
    (e) Not defined

    ANSWER:
    (b)

17. Match the four directions (North, South, East, West) to the direction of $\vec{v}$ under the conditions

    (a) $\vec{v} \cdot \vec{i} > 0$ and $\vec{v} \cdot \vec{j} = 0$
    (b) $\vec{v} \cdot \vec{i} = 0$ and $\vec{v} \cdot \vec{j} > 0$
    (c) $\vec{v} \cdot \vec{i} < 0$ and $\vec{v} \cdot \vec{j} = 0$
    (d) $\vec{v} \cdot \vec{i} = 0$ and $\vec{v} \cdot \vec{j} < 0$

    ANSWER:

    (a) East
    (b) North
    (c) West
    (d) South

ConcepTests and Answers and Comments for Section 10.5

Assume **B** is a 5×5 matrix and $\vec{u} = (1, 2, 3, 4, 5)$. For Problems 1–4, identify the expression as being a
(a) Matrix   (b) Vector   (c) Scalar   (d) Undefined.

1. **B** + **B**
    ANSWER:
    (a)

2. **B** + $\vec{u}$
    ANSWER:
    (d)

3. **B** + $\frac{1}{3}$
    ANSWER:
    (d)

4. $\frac{1}{2}$**B**
    ANSWER:
    (a)

5. If **A** is a 3×4 matrix, it has how many columns?

    (a) 1
    (b) 3
    (c) 4
    (d) 12

    ANSWER:
    (c)

6. If **A** is a 3×4 matrix, it has how many entries?

    (a) 1
    (b) 3
    (c) 4
    (d) 12

    ANSWER:
    (d)

For Problems 7–13, $\mathbf{A}$ and $\mathbf{B}$ are 6×6 matrices and have entries given by $a_{ij} = i+j$ and $b_{ij} = i-j$. Also define $\mathbf{C} = \mathbf{A} + \mathbf{B}$ and $\mathbf{D} = \dfrac{1}{2}\mathbf{C}$. Find the given matrix entry.

7. a_{31}

    (a) 1

    (b) 2

    (c) 3

    (d) 4

    (e) None of the above

       ANSWER:

       (d)

8. b_{14}

    (a) -1

    (b) -3

    (c) 1

    (d) 3

    (e) None of the above

       ANSWER:

       (b)

9. c_{11}

    (a) 1

    (b) 2

    (c) 3

    (d) 4

    (e) None of the above

       ANSWER:

       (b)

10. c_{21}

    (a) 1

    (b) 2

    (c) 3

    (d) 4

    (e) None of the above

       ANSWER:

       (d)

11. c_{12}

    (a) 1

    (b) 2

    (c) 3

    (d) 4

    (e) None of the above

       ANSWER:

       (b)

12. d_{31}

    (a) 1

    (b) 2

    (c) 3

    (d) 4

    (e) None of the above

       ANSWER:

       (c)

13. d_{13}

    (a) 1

    (b) 2

    (c) 3

    (d) 4

    (e) None of the above

      ANSWER:

      (a)

14. The matrix notation, a_{12}, represents

    (a) The entry in row 1 and column 2

    (b) The entry in row 2 and column 1

    (c) The 12th entry

    (d) $a = 12$

    (e) None of the above

      ANSWER:

      (a)

Chapter Eleven

ConcepTests and Answers and Comments for Section 11.1 ━━━━━━━━━━━

1. Find the sixth term of the sequence $2, 4, 6, 8, \ldots$.

   (a) 14
   (b) 12
   (c) 10
   (d) 6
   (e) None of these

   ANSWER:
   (b)

2. Find the sixth term of the sequence $1, 4, 9, 16, \ldots$.

   (a) 18
   (b) 25
   (c) 36
   (d) 49
   (e) None of these

   ANSWER:
   (c)

For Problems 3–11, identify the sequence as
(a) Arithmetic   (b) Geometric   (c) Neither.

3. $2, 4, 6, 8, \ldots$
   ANSWER:
   (a)

4. $1, 4, 9, 16, \ldots$
   ANSWER:
   (c)

5. The sequence defined by $a_n = 3n + 2$
   ANSWER:
   (a)

6. The sequence defined by $a_n = 3^n$
   ANSWER:
   (b)

7. The sequence defined by $a_n = n^2 - 2$
   ANSWER:
   (c)

8. The sequence defined by $a_n = 10^n - 1$
   ANSWER:
   (c)

9. The sequence defined by $a_n = 10^{n-1}$
   ANSWER:
   (b)

10. $6.01, 8.02, 10.03, 12.04, \ldots$
    ANSWER:
    (a)

11. $-1, 1, -2, 2, -3, 3 \ldots$
    ANSWER:
    (c)

For Problems 12–16, identify the statement as being true
(a) Always   (b) Sometimes   (c) Never.

12. If successive terms of an arithmetic sequence change signs, there will be no further sign changes.
    ANSWER:
    (a)

13. If successive terms of a geometric sequence change signs, there will be no further sign changes.
    ANSWER:
    (c)

14. The terms of a geometric sequence get large quickly.
    ANSWER:
    (b)

15. For a geometric sequence, the values of $|a_n|$ are increasing, as $n \to \infty$.
    ANSWER:
    (b)

16. For an arithmetic sequence, the values of $|a_n|$ are increasing, as $n \to \infty$.
    ANSWER:
    (b)
    COMMENT:
    Students may forget that constant sequences are arithmetic.

17. Match the use of the letter a in the four expressions
    (I) a_n (II) a^n (III) $a(n)$ (IV) $a \cdot n$

    (a) A base of an exponential expression
    (b) The name of a function
    (c) A term in a sequence
    (d) A parameter

    ANSWER:

    (a) (II)
    (b) (III)
    (c) (I)
    (d) (IV)

18. If the sequence $-5, -2, 1, 4, 7, \ldots$, is written as a function, its equation is:

    (a) $f(n) = 3n - 2$
    (b) $f(n) = 3n - 8$
    (c) $f(n) = 3n - 5$
    (d) $f(n) = -5 \cdot 3^n$
    (e) None of the above

    ANSWER:
    (b)

19. If the sequence $3, 9, 27, 81, 243, \ldots$, is written as a function, its equation is:

    (a) $f(n) = 5^n - 2$
    (b) $f(n) = 3^n - 2$
    (c) $f(n) = 6 - 3^n$
    (d) $f(n) = 3n$
    (e) None of the above

    ANSWER:
    (e)

ConcepTests and Answers and Comments for Section 11.2 ———————————

1. For the sequence $-1, 2, 4, 3, -2, \ldots$, identify the given equation as true or false.

    (a) $a_3 = 5$
    (b) $a_4 = 3$
    (c) $S_3 = 5$
    (d) $S_1 = a_1$

    ANSWER:

    (a) False
    (b) True
    (c) True
    (d) True

2. For the sequence $a_1, a_2, a_3, \ldots$, the expression $\sum_{i=10}^{20} a_i$ is equivalent to

    (a) S_{20}
    (b) $S_{20} - S_{10}$
    (c) $S_{20} - S_9$
    (d) S_{10}
    (e) None of the above

    ANSWER:

    (c)

3. How many terms are summed in the series $\sum_{i=0}^{20} a_i$?

    (a) a_{20}
    (b) 0
    (c) 20
    (d) 21
    (e) None of the above

    ANSWER:

    (d)

4. True or false? $\sum_{i=1}^{8} 2 = 16$

    ANSWER:
    True

For Problems 5–11, assume the sequences $\{a_i\}$ and $\{b_i\}$ are not all zeros. Identify each statement as being
(a) True   (b) False   (c) Unknown.

5. $\displaystyle\sum_{i=1}^{8} a_i$ is a sequence of eight terms

   ANSWER:

   (b)

6. $\displaystyle\sum_{i=1}^{8}(a_i + b_i) = \sum_{i=1}^{8} a_i + \sum_{i=1}^{8} b_i$

   ANSWER:

   (a)

7. $\displaystyle\sum_{i=1}^{8} a_i$ is an integer

   ANSWER:

   (c)

8. $\displaystyle\sum_{i=0}^{8} a_i = \sum_{k=0}^{8} a_k$

   ANSWER:

   (a)

9. $5 + \displaystyle\sum_{i=1}^{10} a_i = \sum_{i=1}^{10}(5 + a_i)$

   ANSWER:

   (b)

10. $\displaystyle\sum_{i=1}^{5} a_i = \sum_{i=1}^{6} a_i$

    ANSWER:

    (c)

11. $\displaystyle\sum_{i=1}^{10} 5a_i = 5\sum_{i=1}^{10} a_i$

    ANSWER:

    (a)

12. The sum of the first n terms of an arithmetic series is given by $S_n = \frac{1}{2}n$(first term + last term). Identify each statement as true, false or unknown.

    (a) This formula requires n to be even so that $\frac{1}{2}n$ is an integer.
    (b) The first and last terms are integer values.
    (c) If you only know the value of S_8 then you can find the values of the first and last terms.
    (d) If you know the first term, last terms, and the sum, then you can find n.
    (e) The sum formula is only valid for series with $a_1 > 0$.
    (f) The sum of the sequence 1, 4, 9, 16, 25, 36 is $\frac{1}{2}(6)(1 + 36)$.

    ANSWER:

    (a) False
    (b) Unknown
    (c) False
    (d) Unknown, depends if (first + last)=0
    (e) False
    (f) False

ConcepTests and Answers and Comments for Section 11.3 ─────────

For Problems 1–4, determine the requested value if $\displaystyle\sum_{i=0}^{n-1} ar^i = 2^2 + 2^3 + 2^4 + 2^5$.

1. The value of a is

   (a) 1
   (b) 2
   (c) 4
   (d) 5
   (e) None of the above

   ANSWER:
   (c)
   COMMENT:

   There is an assumption that the right-hand side remains a sum of 4 terms and is not considered as $\displaystyle\sum_{i=0}^{n-1} ar^i = 60$.

2. The value of r is

   (a) 1
   (b) 2
   (c) 4
   (d) 5
   (e) None of the above

   ANSWER:
   (b)

3. The value of n is

   (a) 1
   (b) 2
   (c) 4
   (d) 5
   (e) None of the above

   ANSWER:
   (c)

4. The value of the sum is

   (a) $\dfrac{2^2(1-2^3)}{1-2}$

   (b) $\dfrac{2^2(1-2^5)}{1-4}$

   (c) $\dfrac{2^2(1-2^4)}{1-4}$

   (d) $\dfrac{2^2(1-2^4)}{1-2}$

   (e) None of the above

   ANSWER:
   (d)

For Problems 5–8, determine the requested value if $Q_n = \sum_{i=0}^{n-1} ar^i = 2000 + 2000(\frac{1}{5}) + 2000(\frac{1}{5})^2 + 2000(\frac{1}{5})^3 + 2000(\frac{1}{5})^4$.

5. The value of a is

 (a) 1/5
 (b) 5
 (c) 1000
 (d) 2000
 (e) None of the above

   ANSWER:
   (d)

6. The value of r is

 (a) 1
 (b) 1/5
 (c) 5
 (d) 5000
 (e) None of the above

   ANSWER:
   (b)

7. The value of n is

 (a) 5
 (b) 4
 (c) 1/5
 (d) 1
 (e) None of the above

   ANSWER:
   (a)

8. The value of the sum is

 (a) $\dfrac{2000(1 - \frac{1}{5})^5}{(1 - \frac{1}{5})}$

 (b) $\dfrac{2000(1 - \frac{1}{5})^4}{(1 - \frac{1}{5})}$

 (c) $\dfrac{2000(1 - \frac{1}{5})^4}{(\frac{1}{5})}$

 (d) $\dfrac{2000(1 - \frac{1}{5})^5}{(1 + \frac{1}{5})}$

 (e) None of the above

   ANSWER:
   (e)

For Problems 9–12, determine the requested value using $T_n = \sum_{i=0}^{n-1} ar^i = -10 + \dfrac{10}{2} - \dfrac{10}{4} + \dfrac{10}{8}$.

9. The value of a is

   (a) 1
   (b) 10
   (c) -1
   (d) -10
   (e) None of the above

   ANSWER:
   (d)

10. The value of r is

    (a) -1
    (b) -1/2
    (c) 1/2
    (d) -10
    (e) None of the above

    ANSWER:
    (b)

11. The value of n is

    (a) 3
    (b) 4
    (c) 5
    (d) 10
    (e) None of the above

    ANSWER:
    (b)

12. The value of the sum is

    (a) $\dfrac{-10(1 - (-\frac{1}{2})^4)}{\frac{3}{2}}$

    (b) $\dfrac{-10(1 - (\frac{1}{2})^4)}{\frac{1}{2}}$

    (c) $\dfrac{-10(1 - (-\frac{1}{2})^4)}{\frac{1}{2}}$

    (d) $\dfrac{-10(1 - (\frac{1}{2})^4)}{\frac{3}{2}}$

    (e) None of the above

    ANSWER:
    (a)

13. The finite sum formula, $S_n = \dfrac{a(1 - r^n)}{1 - r}$, is applicable to which of the following sequences

    (a) $2, 4, 6, 8$
    (b) $a_n = 2n - 3$
    (c) $1, \frac{1}{5}, (\frac{1}{5})^2, (\frac{1}{5})^3$
    (d) $a_n = -4.1(-\frac{1}{5})^n$
    (e) All of the above
    (f) None of the above

    ANSWER:
    (c) and (d)

14. Does $\sum_{i=0}^{n-1} ar^i = \sum_{i=1}^{n} ar^{i-1}$?

    ANSWER:
    Yes

ConcepTests and Answers and Comments for Section 11.4

1. In the sigma notation, the series $S = 2^2 + 2^3 + 2^4 + \ldots$ is

   (a) $\displaystyle\sum_{i=0}^{\infty} 2^{i+2}$

   (b) $\displaystyle\sum_{i=2}^{\infty} 2^i$

   (c) $\displaystyle\sum_{k=1}^{\infty} 2^{k+1}$

   (d) All of the above

   (e) None of the above

   ANSWER:

   (d)

2. The expression $\displaystyle\sum_{i=1}^{100} a_i - \sum_{i=0}^{100} a_i$ can be simplified to

   (a) $\displaystyle -\sum_{i=1}^{100} a_i$

   (b) $-a_0$

   (c) $\displaystyle\sum_{k=1}^{100} a_{k+1}$

   (d) $\displaystyle\sum_{i=1}^{100} (a_{i-1} - a_i)$

   (e) Cannot be simplified

   ANSWER:

   (b)

3. The expression $\displaystyle\sum_{i=1}^{50} a_i + \sum_{i=0}^{50} a_i$ can be simplified to

   (a) $\displaystyle a_0 + 2\sum_{i=1}^{50} a_i$

   (b) $\displaystyle\sum_{i=0}^{50} (a_i + a_{i+1})$

   (c) $\displaystyle a_0 + \sum_{i=1}^{50} 2a_i$

   (d) All of the above

   (e) None of the above

   ANSWER:

   (a) and (c)

4. The expression $\displaystyle\sum_{i=0}^{\infty} 5(0.10)^i$ has a value of

   (a) $50/9$

   (b) $45/10$

   (c) 50

   (d) ∞

   (e) None of the above

   ANSWER:

   (a)

5. If $\sum\limits_{i=0}^{\infty} k(0.50)^i = 5$, then k has a value of

   (a) 0.5

   (b) 2.5

   (c) 5

   (d) 10

   (e) None of the above

      ANSWER:

      (b)

6. If $\sum\limits_{i=0}^{\infty} 10p^i = 8$, then p has a value of

   (a) 0.25

   (b) 2.5

   (c) −0.25

   (d) −2.5

   (e) None of the above

      ANSWER:

      (c)

7. The expression $\sum\limits_{i=0}^{\infty} 10(-1)^i$ has a value of

   (a) 0

   (b) 1

   (c) $-\infty$

   (d) ∞

   (e) None of the above

      ANSWER:

      (e)

8. The expression $\sum\limits_{i=0}^{\infty} 2^i$ has a value of

   (a) 0

   (b) 1

   (c) 2

   (d) $1/(1 - 2)$

   (e) None of the above

      ANSWER:

      (e)

9. The expression $\sum\limits_{i=0}^{\infty} (-0.2)^i$ has a value of

   (a) 0

   (b) 6/5

   (c) 5/6

   (d) $-\infty$

   (e) None of the above

      ANSWER:

      (c)

10. The present value of 20 annual lottery payments is

    (a) More than the total payments
    (b) Less than the total payments
    (c) Equal to the total payments
    (d) None of the above

    ANSWER:

    (b)

    COMMENT:

    There could be discussion about the interest rate. We assume it is greater than zero. However if it were zero the answer would change to (c).

11. If payment of an amount is put off forever, the present value of the amount is

    (a) Zero
    (b) More than zero, but less than the amount
    (c) Equal to the amount
    (d) More than the amount
    (e) None of the above

    ANSWER:

    (a)

12. A store offers a $1000 laptop with a delayed payment for one year. If you consider your rate of return on investment to be 5%, then the present value in dollars of the offer is

    (a) 1000
    (b) 1000(1.05)
    (c) 1000/(1.05)
    (d) 1000-1000/1.05
    (e) None of the above

    ANSWER:

    (d)

Chapter Twelve

Chapter Twelve

ConcepTests and Answers and Comments for Section 12.1

1. Describe the graph derived from the following parametric formulas.

$$\begin{cases} x = t + 1 & y = 0 & \text{if} \quad 0 < t \le 1 \\ x = 2 & y = 2(t-1) & \text{if} \quad 1 < t \le 2 \\ x = 4 - t & y = 2 & \text{if} \quad 2 < t \le 3 \\ x = 1 & y = 2(4-t) & \text{if} \quad 3 < t \le 4 \end{cases}$$

(a) Function
(b) Square
(c) Diamond
(d) Rectangle
(e) None of the above

ANSWER:
(d)

2. Without using a graphing utility, identify how many of the intervals, (i) to (iv), give the same parametric graph as $x = \cos t, y = \sin t, 0 \le t \le 4\pi$

(i) $0 \le t \le 2\pi$
(ii) $-\pi \le t \le 2\pi$
(iii) $0 \le t \le 6\pi$
(iv) $2\pi \le t < 4\pi$.

(a) None
(b) One
(c) Two
(d) Three
(e) Four

ANSWER:
(e)

3. Without using a graphing utility, determine which graphs of the parametric equation match the graph of $x = \cos t$, $y = \sin t$ for $0 \le t \le 2\pi$.

(a) $x = \cos \frac{1}{2}t, y = \sin \frac{1}{2}t$ for $0 \le t \le 2\pi$
(b) $x = \cos 2t, y = \sin 2t$ for $0 \le t \le 2\pi$
(c) $x = \frac{1}{2}\cos t, y = \frac{1}{2}\sin t$ for $0 \le t \le 2\pi$
(d) $x = \sin t, y = \cos t$ for $0 < t < 2\pi$
(e) $x = -\cos t, y = \sin t$ for $0 \le t \le 2\pi$

ANSWER:

(a) No
(b) Yes
(c) No
(d) No
(e) Yes

COMMENT:
The graph does not show all the information in the algebraic definition. For example, it does not distinguish between Part (b) and (e).

4. Which equation produces the graph of a circle of radius 3, centered at $(2, 1)$?

   (a) $(x - 2)^2 + (y - 1)^2 = 9$,

   (b) $r = 3, 0 \leq \theta < 2\pi$

   (c) $x = 2 + \cos t, y = 1 + \sin t$ for $0 \leq t \leq 2\pi$

   (d) All of the above

   (e) None of the above

   > ANSWER:
   >
   > (a)

5. Which of the following equations has a diamond shaped graph?

   (a) $\begin{cases} x = t & y = 1 - |t| & \text{if} & -1 \leq t \leq 1 \\ x = t - 2 & y = |t - 2| - 1 & \text{if} & 1 \leq t \leq 3 \end{cases}$

   (b) $\begin{cases} x = t & y = t - 1 & \text{if} & 0 < t \leq 1 \\ x = 2 - t & y = t - 1 & \text{if} & 1 < t \leq 2 \\ x = 2 - t & y = t - 3 & \text{if} & 2 < t \leq 3 \\ x = 3 - t & y = 4 - t & \text{if} & 3 < t \leq 4 \end{cases}$

   (c) $\begin{cases} x = t & y = 1 - t & \text{if} & 0 < t \leq 1 \\ x = 2 - t & y = 1 - t & \text{if} & 1 < t \leq 2 \\ x = 2 - t & y = t - 3 & \text{if} & 2 < t \leq 3 \\ x = t - 4 & y = t - 3 & \text{if} & 3 < t \leq 4 \end{cases}$

   (d) All of the above

   (e) None of the above

   > ANSWER:
   >
   > (d)

For Problems 6–11, identify the parametric graph on the interval $0 \leq t \leq 1$ as

(a) Horizontal, left to right

(b) Horizontal, right to left

(c) Vertical, up

(d) Vertical, down

(e) None of the above

6. $x = 4, y = t - 3$

   > ANSWER:
   >
   > (c)

7. $x = t, y = t$

   > ANSWER:
   >
   > (e)

8. $x = 0, y = t^2$

   > ANSWER:
   >
   > (c)

9. $x = 4t, y = -3$

   > ANSWER:
   >
   > (a)

10. $x = 1 - t, y = 1$

    > ANSWER:
    >
    > (b)

11. $y = 1 - t^2, x = -2$

    > ANSWER:
    >
    > (d), Note the interchange of x and y.

For Problems 12–16, compare the graph of the given equation to that of $x = t$, $y = 1 - |t|$ for $-1 \leq t \leq 1$. Note: the same motion means the same direction and velocity.

(a) Same graph and motion

(b) Same graph, different motion

(c) Different graph

12. $x = -t$, $y = 1 - |t|$ for $-1 \leq t \leq 1$
    ANSWER:
    (b)

13. $x = t$, $y = |t| - 1$ for $-1 \leq t \leq 1$
    ANSWER:
    (c)

14. $x = t$, $y = 1 - t^2$ for $-1 \leq t \leq 1$
    ANSWER:
    (c)

15. $x = \frac{1}{2}t$, $y = 1 - |\frac{1}{2}t|$ for $-2 \leq t \leq 2$
    ANSWER:
    (b)

16. $x = t$, $y = 2 - |t|$ for $-1 \leq t \leq 1$
    ANSWER:
    (c)

ConcepTests and Answers and Comments for Section 12.2 ────────────

1. When graphed, which of the following equations produce a circle?

   (a) $5y^2 = 1 - 5x^2$
   (b) $2y^2 + 3x^2 - 1 = 0$
   (c) $2y^2 = 3x^2 + 1$
   (d) $5y = 5x^2 + 1$
   (e) None of the above

   ANSWER:
   (a)

2. When graphed, which of the following equations produce a circle?

   (a) $2 + 3y^2 = 5 + 3x^2$
   (b) $4y^2 + 4x^2 + 1 = 0$
   (c) $2y^2 - 1 = 3 - 2x^2$
   (d) $5y + 5x = 1$
   (e) None of the above

   ANSWER:
   (c)

3. When graphed, which of the following equations produce a circle?

   (a) $(x + 1)^2 + 3(y + 1)^2 = 5$
   (b) $y^2 + (x - 1)(x + 1) = 0$
   (c) $3y^2 = 3 - 2(x + 3)^2$
   (d) $5y^3 + 5x^3 = 8$
   (e) None of the above

   ANSWER:
   (b)

4. When graphed, which of the following equations produce a circle?

   (a) $x = \cos(\pi t), y = \sin(\pi t)$ for $0 \leq t \leq 2$
   (b) $x = \sin t, y = \cos t$ for $0 \leq t \leq 2\pi$
   (c) $x = -\cos(2\pi t), y = \sin(2\pi t)$ for $0 \leq t \leq 1$
   (d) All of the above
   (e) None of the above

   ANSWER:
   (d)

5. When graphed, which of the following equations produce a circle?

   (a) $x = \cos(2\pi t), y = 1 + \sin(2\pi t)$ for $0 \leq t \leq 2$
   (b) $x = \cos \frac{1}{2}t, y = \sin \frac{1}{2}t$ for $0 \leq t \leq 2\pi$
   (c) $x = \cos(2\pi t), y = \sin(4\pi t)$ for $0 \leq t \leq 1$
   (d) All of the above
   (e) None of the above

   ANSWER:
   (a)

6. When graphed, which of the following equations produce a circle?

   (a) $x = 1 - \cos(t - 1), y = \sin(t - 2)$ for $0 \leq t \leq 2\pi$
   (b) $x = 2 + \cos \frac{1}{2}t, y = 2 \sin \frac{1}{2}t$ for $0 \leq t \leq 4\pi$
   (c) $x = 1 - \cos t^2, y = 3 + \sin t^2$ for $0 \leq t \leq 2\pi$
   (d) All of the above
   (e) None of the above

   ANSWER:
   (c)

7. When graphed, which of the following equations produces the circle shown in Figure 12.1?

   (a) $x = 2 + 3\cos t, y = 3 + 3\sin t$
   (b) $x = 2 + 2\cos t, y = 3 + 2\sin t$
   (c) $x = 3 + 2\cos t, y = 2 + 2\sin t$
   (d) $x = 2 + \cos 2t, y = 3 + 3\sin 2t$
   (e) Cannot be determined

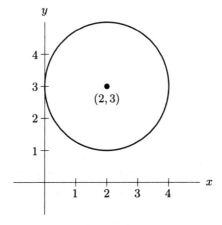

Figure 12.1

ANSWER:
(b)

8. When graphed, which of the following equations produce the circle shown in Figure 12.1?

    (a) $(x - 2)^2 + (y - 3)^2 = 2$
    (b) $(x + 2)^2 + (y + 3)^2 = 4$
    (c) $(x - 2)^2 + (y - 3)^2 = 4$
    (d) $(x - 2)^2 + (y - 3)^2 = 1$
    (e) Cannot be determined

    ANSWER:
    (c)

9. Give the center and radius of the circle whose equation is $(x - a)^2 + (y - b)^2 = c$.

    (a) $(a, b), c$
    (b) $(a, b), \sqrt{c}$
    (c) $(b, a), c$
    (d) $(b, a), \sqrt{c}$
    (e) None of the above

    ANSWER:
    (b)

10. Give the center and radius of the circle whose equation is $x = a + b \cos t, y = c + b \sin t$, for $0 \leq t \leq 2\pi$. Assume $b > 0$.

    (a) $(a, b), c$
    (b) $(a, c), \sqrt{b}$
    (c) $(a, c), b$
    (d) $(a, c), b^2$
    (e) None of the above

    ANSWER:
    (c)

11. For each equation in parts (a) - (g) state whether y is an explicit function of x.

    (a) $x = y$
    (b) $x = 2$
    (c) $y = 4$
    (d) $(x - 2)^2 = y - 3$
    (e) $y = |x|$
    (f) $y^2 = 3x$
    (g) $y^2 = 1 - x^2$

        ANSWER:

    (a) no
    (b) no
    (c) yes
    (d) no
    (e) yes
    (f) no
    (g) no

ConcepTests and Answers and Comments for Section 12.3 ▬▬▬▬▬

1. Which of the following equations graph an ellipse that is not a circle?

    (a) $5y^2 = 1 - 5x^2$
    (b) $2y^2 + 3x^2 - 1 = 0$
    (c) $2y^2 = 3x^2 + 1$
    (d) $5y = 5x^2 + 1$
    (e) None of the above

    ANSWER:
    (b)

2. Which of the following equations graph an ellipse that is not a circle?

   (a) $3y^2 = 5 + 2x^2$
   (b) $4y^2 + 3x^2 + 1 = 0$
   (c) $5 + 2y^2 = 3 - 2x^2$
   (d) $5x^2 + y^2 = 7$
   (e) None of the above

   ANSWER:
   (d)

3. Which of the following equations graph an ellipse that is not a circle?

   (a) $(x + 1)^2 + 3(y + 1)^2 = 5$
   (b) $y^2 + (x - 1)(x + 1) = 0$
   (c) $3x^2 = 3 + 2(y - 3)^2$
   (d) $(x - 3)^3 + 5(y - 2)^3 = 8$
   (e) None of the above

   ANSWER:
   (a)

4. Which of the following equations graph an ellipse that is not a circle?

   (a) $x = \cos(3\pi t), y = -\sin(2\pi t)$ for $0 \leq t \leq 2$
   (b) $x = 2 + \sin t, y = 3 + \cos t$ for $0 \leq t \leq 2\pi$
   (c) $x = -3\cos(2\pi t), y = \sin(2\pi t)$ for $0 \leq t \leq 1$
   (d) All of the above
   (e) None of the above

   ANSWER:
   (c)

5. Which of the following equations graph an ellipse that is not a circle?

   (a) $x = 3 + \cos(2\pi t), y = 2 + \sin(2\pi t)$ for $0 \leq t \leq 2$
   (b) $x = 3\cos 2t, y = 3 + 5\sin 2t$ for $0 \leq t \leq 2\pi$
   (c) $x = 1 + \cos(2\pi t), y = 2 - 3\sin(4\pi t)$ for $0 \leq t \leq 1$
   (d) All of the above
   (e) None of the above

   ANSWER:
   (b)

6. Which of the following equations graph an ellipse that is not a circle?

   (a) $x = \sin(t - 1), y = 3\cos(t - 1)$ for $0 \leq t \leq 2\pi$
   (b) $x = 3\cos t^2, y = 5\sin t^2$ for $0 \leq t \leq 2\pi$
   (c) $x = 5 + \cos \frac{1}{2}t, y = 5\sin \frac{1}{2}t$ for $0 \leq t \leq 4\pi$
   (d) All of the above
   (e) None of the above

   ANSWER:
   (d)

7. When graphed, which of the following equations produce the ellipse shown in Figure 12.2?

   (a) $x = 2 + \cos t, y = 3 + 2\sin t$
   (b) $x = 3 + 2\cos t, y = 2 + \sin t$
   (c) $x = 3 + \cos t, y = 2 + 2\sin t$
   (d) $x = 3 + \cos t, y = 2 + \sin 2t$
   (e) Cannot be determined

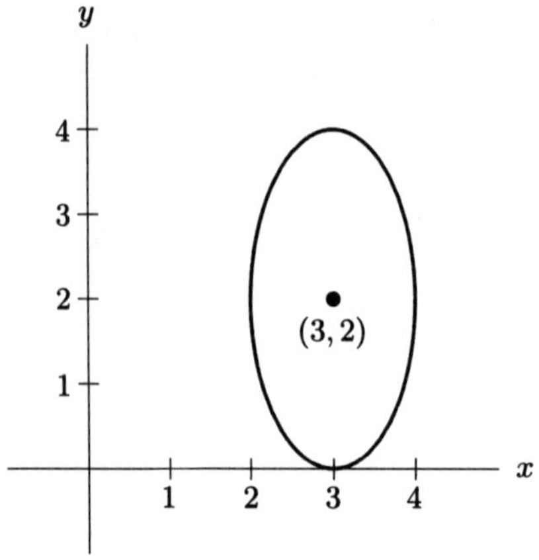

Figure 12.2

ANSWER:

(c)

8. When graphed, which of the following equations produce the ellipse shown in Figure 12.2?

    (a) $(x - 3)^2 + \dfrac{(y - 2)^2}{2} = 1$
    (b) $4(x + 3)^2 + (y + 2)^2 = 4$
    (c) $(x - 3)^2 + 4(y - 2)^2 = 1$
    (d) $4(x - 3)^2 + (y - 2)^2 = 4$
    (e) Cannot be determined

    ANSWER:

    (d)

9. Give the lengths of the horizontal and vertical axis of the ellipse produced by the equation $4x^2 + y^2 = c$.

    (a) $c, 2c$
    (b) $\sqrt{c}, 2\sqrt{c}$
    (c) $\sqrt{c}, \sqrt{c}/2$
    (d) $2\sqrt{c}, \sqrt{c}$
    (e) None of the above

    ANSWER:

    (b)

10. Give the lengths of the horizontal and vertical axis of the ellipse produced by the equation $x = a + b\cos t, y = c + d\sin t$.

    (a) $2a, 2c$
    (b) b, d
    (c) $2b, 2d$
    (d) $b/2, c/2$
    (e) None of the above

    ANSWER:

    (c)

ConcepTests and Answers and Comments for Section 12.4 ———

1. When graphed, which of the following equations produce the hyperbola shown in Figure 12.3?

    (a) $x = 3 + \sec t, y = 3 + 2\tan t$
    (b) $x = 3 + \tan t, y = 2 + 2\sec t$
    (c) $x = 3 + \tan t, y = 2 + \sec t$
    (d) $x = 3 + 2\tan t, y = 2 + \sec t$
    (e) Cannot be determined

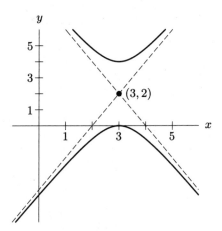

Figure 12.3

   ANSWER:
   (b)

2. When graphed, which of the following implicit equations produce the hyperbola shown in Figure 12.3?

    (a) $\dfrac{(x-3)^2}{1} - \dfrac{(y-2)^2}{4} = 1$
    (b) $\dfrac{(x-3)^2}{4} - \dfrac{(y-2)^2}{1} = 1$
    (c) $\dfrac{-(x-3)^2}{1} + \dfrac{(y-2)^2}{4} = 1$
    (d) $\dfrac{-(x-3)^2}{4} + \dfrac{(y-2)^2}{1} = 1$
    (e) Cannot be determined

   ANSWER:
   (d)

3. The unit hyperbola has the equation $x^2 - y^2 = 1$. What is this equation in parametric form for $0 \le t \le 2\pi$?

    (a) $x = \cos t, y = \sin t$
    (b) $x = \tan t, y = \sec t$
    (c) $x = \sec t, y = \tan t$
    (d) $x = 1 + \sec t, y = 1 + \tan t$
    (e) None of the above

   ANSWER:
   (c)

4. If the dotted box in Figure 12.4 is centered at $(5, 4)$ and has a width of 4 and a height of 6, what is the equation of the hyperbola?

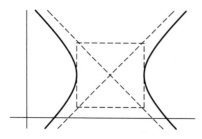

Figure 12.4

(a) $\dfrac{(x-5)^2}{4} + \dfrac{(y-4)^2}{9} = 1$

(b) $\dfrac{(x-5)^2}{4} - \dfrac{(y-4)^2}{6} = 1$

(c) $-\dfrac{(x-5)^2}{9} + \dfrac{(y-4)^2}{4} = 1$

(d) $\dfrac{(x-5)^2}{4} - \dfrac{(y-4)^2}{9} = 1$

(e) Cannot be determined

ANSWER:

(d)

For Problems 5–17, identify the given equation as producing one of the following conic sections:

(a) Parabola

(b) Circle

(c) Ellipse, that is not a circle

(d) Hyperbola

(e) None of the above

5. $3y^2 = 5 + 2x^2$

   ANSWER:

   (d)

6. $4y^2 + 4x^2 + 1 = 0$

   ANSWER:

   (e)

7. $2y^2 = 3 - 2x^2$

   ANSWER:

   (b)

8. $5x^2 + y^2 = 7$

   ANSWER:

   (c)

9. $x = 3 + \cos(2\pi t), y = 2 + \sin(2\pi t)$ for $0 \le t \le 2$

   ANSWER:

   (b)

10. $x = 3\cos 2t, y = 3 + 5\sin 2t$ for $0 \le t \le 2\pi$

    ANSWER:

    (c)

11. $x = 1 + \cos(2\pi t), y = 2 - 3\sin(4\pi t)$ for $0 \le t \le 1$

    ANSWER:

    (e)

12. $x = \sin(t - 1), y = 3\cos(t - 1)$ for $0 \le t \le 2\pi$

    ANSWER:

    (c)

13. $x = -\sin t, y = \cos t$ for $0 \le t \le 2\pi$

    ANSWER:

    (b)

14. $x = t, y = t^2$ for $-\infty < t < \infty$

    ANSWER:

    (a)

15. $5y^2 = 1 - 5x^2$

    ANSWER:

    (b)

16. $2y^2 + 3x^2 - 1 = 0$

    ANSWER:

    (c)

17. $2y^2 = 3x^2 + 1$

    ANSWER:

    (d)

For Problems 18–21, the graph of the hyperbola in Figure 12.5 has the equation $x = 5 + 2\sec t$, $y = 4 + 3\tan t$ for $0 \leq t \leq 2\pi$.

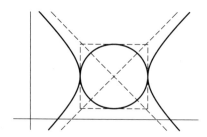

Figure 12.5

18. What is the equation of the ellipse inside the box?

    (a) $(x - 5)^2 + (y - 4)^2 = 4$
    (b) $x = 5 + 3\sin t, y = 4 + 2\cos t$
    (c) $x = 5 + 2\cos t, y = 4 + 3\sin t$
    (d) $x = 4 + 3\sin t, y = 5 + 2\cos t$
    (e) Cannot be determined

    ANSWER:
    (c)

19. What are the coordinates of the center of the box?

    (a) $(5, 4)$
    (b) $(4, 5)$
    (c) $(2, 3)$
    (d) $(3, 2)$
    (e) Cannot be determined

    ANSWER:
    (a)

20. What is the width of the box?

    (a) 2
    (b) 4
    (c) 5
    (d) 6
    (e) Cannot be determined

    ANSWER:
    (b)

21. What is the height of the box?

    (a) 2
    (b) 4
    (c) 5
    (d) 6
    (e) Cannot be determined

    ANSWER:
    (d)

ConcepTests and Answers and Comments for Section 12.5 ━━━━━━━━━

1. There is only one ellipse with two given focal points.

    (a) True
    (b) False

    ANSWER:
    (b). False.
    COMMENT:
    Ask the students what information in addition to the locations of the focal points would be sufficient to determine a unique ellipse.

2. In the morning you walk from one focus of an ellipse to the other focus by first walking directly from the first focus to a point A on the ellipse and then turning to walk directly to the second focus. On the afternoon return trip, you go by way of a different point B on the ellipse. Do you walk the same distance on both trips?

    (a) Yes
    (b) No

    ANSWER:
    (a). Yes, the geometric definition of an ellipse is that the lengths of the paths do not depend at all on the particular points A and B.

3. The directrix of a parabola is an asymptote of the parabola.

    (a) True
    (b) False

    ANSWER:
    (b). False. Parabolas do not have asymptotes.
    COMMENT:
    The directrix is a line perpendicular to the line of symmetry of the parabola.

ConcepTests and Answers and Comments for Section 12.6

1. The point $(\cosh 5, \sinh 5)$ is on the hyperbola $x^2 - y^2 = 1$.

   (a) True
   (b) False

   ANSWER:
   (a). True.
   COMMENT:
   The points $(\cosh t, \sinh t)$ are on the hyperbola $x^2 - y^2 = 1$ for all t. Ask if all points on the hyperbola are of the form $(\cosh t, \sinh t)$. They are not. Only the points on one branch of the hyperbola are of this form.

2. The values of $\cosh x$ and $\sinh x$ are very close for large x.

   (a) True
   (b) False

   ANSWER:
   (a). True. For large x, we have

   $$\cosh x = \frac{e^x + e^{-x}}{2} \approx \frac{e^x + 0}{2} = \frac{1}{2}e^x$$

   and

   $$\sinh x = \frac{e^x - e^{-x}}{2} \approx \frac{e^x - 0}{2} = \frac{1}{2}e^x$$

   COMMENT:
   Ask the students about the difference $\cosh x - \sinh x$. It equals e^{-x}.

3. The graph of $y = \tanh x$ has the horizontal asymptote $y = 0$ as $x \to \infty$.

   (a) True
   (b) False

   ANSWER:
   (b). False. Since

   $$\tanh x = \frac{\sinh x}{\cosh x} = \frac{e^x - e^{-x}}{e^x + e^{-x}} = \frac{1 - e^{-2x}}{1 + e^{-2x}},$$

   we have

   $$\tanh x \to 1 \text{ as } x \to \infty.$$

   The line $y = 1$ is a horizontal asymptote as $x \to \infty$.
   COMMENT:
   What is the horizontal asymptote as $x \to -\infty$? Ask students to relate the horizontal asymptotes of the graph of $y = \tanh x$ to to asymptotes of the hyperbola $x^2 - y^2 = 1$.

4. The functions $\cosh x$, $\sinh x$, and $\tanh x$ have the same domains.

   (a) True
   (b) False

   ANSWER:
   (b). False. The functions $\cosh x$ and $\sinh x$ are defined for all values of x, but $\tanh x$ is not defined for $x = 0$.

Notes

Notes

<u>Notes</u>

Notes

Notes

Notes

Notes

Notes

Notes

Notes

Notes

Notes

Notes

Notes